Customer Relationship Management in Agri-Business:
A Case of Agri-Business Retailers

Customer Relationship Management in Agri-Business:

A Case of Agri-Business Retailers

Sajad Khurshid Khan
Ex-Scholar, SKUAST-Jammu
Manager, Seagull Const. Company
Banihal Chamalwas, J&K - 182 146
Jammu and Kashmir State

Jyoti Kachroo
Prof. & Head
Division of Agricultural Economics & ABM,
SKUAST-Jammu, Chatha - 180 009
Jammu and Kashmir State

Naveed Hamid
Assistant Professor, Agri-Business Management
Division of Agricultural Economics & ABM,
SKUAST-Jammu, Chatha - 180 009
Jammu and Kashmir State

SSPH

SATISH SERIAL PUBLISHING HOUSE
403, Express Tower, Commercial Complex
Azadpur, Delhi-110033 (India)
Phone : 011-27672852, Fax : 91-11-27672046
e-mail : info@satishserial.com, hkjain1975@yahoo.com
Website : www.satishserial.com

Published by :

SATISH SERIAL PUBLISHING HOUSE
403, Express Tower, Commercial Complex, Azadpur, Delhi-110033 (INDIA)
Phone : 011-27672852 Fax : 91-11-27672046
E-mail : info@satishserial.com, hkjain1975@yahoo.com

ISBN 978-93-86200-63-1

Composed, Designed & Printed in India

Preface

In today's global and competitive economy there is high competition and customers have more and better choices than ever before, therefore, it is difficult to retain customers and create customer loyalty. The agricultural business ventures will be successful only if they adopt a management approach that brings the customers so close to the company's processes. As customer loyalty increasingly becomes vital and indispensible for the successiveness of any business venture.

As customer loyalty increasingly becomes indispensible and important for the agri-businesses and in general to the business ventures, customer relationship management has become a limelight and a new tool. Companies have been increasingly implementing CRM systems. According to Gartner's report 2014, worldwide CRM software market grew 13.7 percent in 2013. As technologies and customer expectations rapidly change, businesses realize the value of having long term relationship with individual customers and other business partners. First chapter provides an overview about the status and importance of agri business/Food sector nationally/state level and customer relationship management concept. Second chapter deals with the past studies related to the customer relationship in agri business followed by third chapter that focuses on importance, objectives, need and results of the research study. Fourth chapter presents the summary/conclusion and policy implications related to the study.

We felt our heartfelt gratitude to Dr. Dileep Kachroo (Registrar SKUAST-J), whose genuine guidance and conceptual technical ideas during the course of planning and timely help in finalizing this book.

Authors

Contents

Chapter 1

Food Sector: An Overview

The India food processing sector is currently growing at an average rate of 13.5 per cent annually. The industry is expecting a higher growth rate due to the government support being provided in terms of improving the supply chain, by building large primary collection and distribution centres across India involving the private sector, in the period 2012-17. Indian Biscuits industry consists of two large scale manufacturers, around 50 medium scale brands and small scale units ranging up to 2500 units. Also, the unorganized sector is estimated to have approximately 30,000 small & tiny bakeries across the country.

Over the past five years (2009-10 to 2013-14), the Indian Biscuits Market has been growing at a rapid pace and is expected to sustain its growth in the coming years. The strong economic growth of the last few years and increasing brand consciousness among the new wealthy Indian middle class combined with a greater exposure to western lifestyle has created hunger for branded biscuits in the country. The Indian food industry has grown from strength to strength in the past few years with recent upswing in the packaged and processed food industry.

Indian bakery industry is one of the biggest sections in the processed food industry of the nation. The rising foreign influence, emergence of female working population and changing consumption patterns have attributed strongly in the growth trajectory of the bakery industry. Indian bakery industry has undergone a massive change majorly on account of changing perception of bakery products and evolving consumer tastes. Bakery products are items of mass consumption due to their affordability and nutrition content.

Trends in the Industry

E-retailing of bakery products

- Expanding foothold
- Improved Packaging
- Innovation in ingredients

Rising urbanization and growth in the disposable incomes of the Indian population has proven to be a magnet for international bakery chains. India has seen an influx of foreign bakery companies foraying into India which in turn has helped in improving the quality of Indian bakery products. Today there is a constant effort by the bakery players to innovate their product line to match up to Indian palate. Driven by evolving perception of bakery products in India and changing consumer preferences, Indian bakery industry will touch levels of Rs. 483 billion in the next five years.

The Indian food industry has grown from strength to strength in the past few years with recent upswing in the packaged and processed food industry. Indian bakery industry is one of the biggest sections in the processed food industry of the nation. The rising foreign influence, emergence of female working population and changing consumption patterns have attributed strongly in the growth trajectory of the bakery industry. Global bakery industry grew at an annual rate of approximately 6.5 per cent during 2007-13 and leading research associations project the global industry to grow at a compound annual growth rate (CAGR) of approximately 7 per cent in the next few years. The global bakery products market is forecasted to reach US$447 billion by the year 2017, driven by new consumer preferences for conveniently portioned and easy-to-consume bakery goods. Bread and Biscuits are fast and convenient food based on wheat. Its consumption is more in cities and towns where industrial working group population is more. Bread and Biscuit consumption is increasing day by day and these are being increasingly used for various feeding programmed for children managed by voluntary agencies and State departments of health. Biscuits, which constitute an important item of bakery industry, have now become a common item of consumption among all classes of people. With tea or coffee, a biscuit makes a tasty and nutritious snack. It has become more popular as a convenient food.

With the changing socio-economic environment i.e. increased literacy rate (about 65 per cent), higher per capita income, larger number of women going out for work to sustain family, higher living standard and increased tourist population etc., one can expect at least 5 per cent growth rate in case of bakery products in coming years. It has been estimated, there are about 1 lakh small scale industries (SSI) units producing items worth Rs.3000 crores annually. The per capita consumption of bakery products is the lowest in the world i.e. 1.27 kg, whereas, in advanced countries it ranges from 50-70 kgs (Bakery Industry in India, 2012). The Indian biscuit sector is dominated by players like Britannia, Parle and Sunfeast brand (ITC) together with other small players like Priyagold, Anmol Biscuits, Cremica etc.

Indian biscuit industry is one of the biggest sections in the processed food industry of the nation and has undergone a massive change majorly on account of changing perception of biscuit products and evolving consumer tastes. Rising urbanization and growth in the disposable incomes of the Indian population has proven to be a magnet for international biscuit chains owing to which the sector has seen an influx of foreign biscuit companies foraying into India which has helped in improving the quality of Indian biscuit products. Today there is a constant effort by the biscuit players to innovate their product line to match up to Indian palate. Driven by evolving perception of bakery products in India, consumption boom in the nation and changing consumer preferences, it is estimated that the Indian bakery industry to touch levels of Rs. 483 billion in the next five years (Bakery Industry in India, 2012).

The biscuit industry in India is gearing up aggressively to tap the medium and premium segment within the country in the coming years. Biscuit industry is witnessing a major shift towards 'premiumisation' as consumer preferences change, fuelled by soaring disposable incomes in smaller towns and health awareness. People with lower incomes are also upgrading from the 'affordable' glucose biscuits to mass cream biscuits and mass cookies. The biscuit industry, which consists of economy, middle, premium and super premium segments, is likely to grow at 14-15 percent annually for the next five years. But the premium and super-premium segments, according to industry experts, are believed to be growing at a much faster rate, probably more than 20 percent on a year on- year basis. Currently, economy and middle segments (mass segment) comprise about 70 percent of the total biscuit market in India.

While biscuits priced below Rs.100 a kg are in the economy segment and biscuits priced at Rs.100-150 per kg are categorised as middle segment. Whereas, biscuits with a price of Rs. 125-150 per kg fall in the premium category and above Rs. 150 a kg in the super-premium segment. Over the last few years, the biggest shift that has been happening (in the country's biscuit industry) is the premiumisation shift and this is best borne out by the reducing contribution of the glucose segment. The glucose segment used to contribute 33 percent by value a couple of years ago and currently this is expected to drop to around 24 percent. Also more action is expected in the health and wellness segment, due to increasing concern about lifestyle-related ailments. There will be more products on the market which are naturally healthy. Many biscuits, which are good for health, but are not advertised as such, will see a change in their marketing pitch.

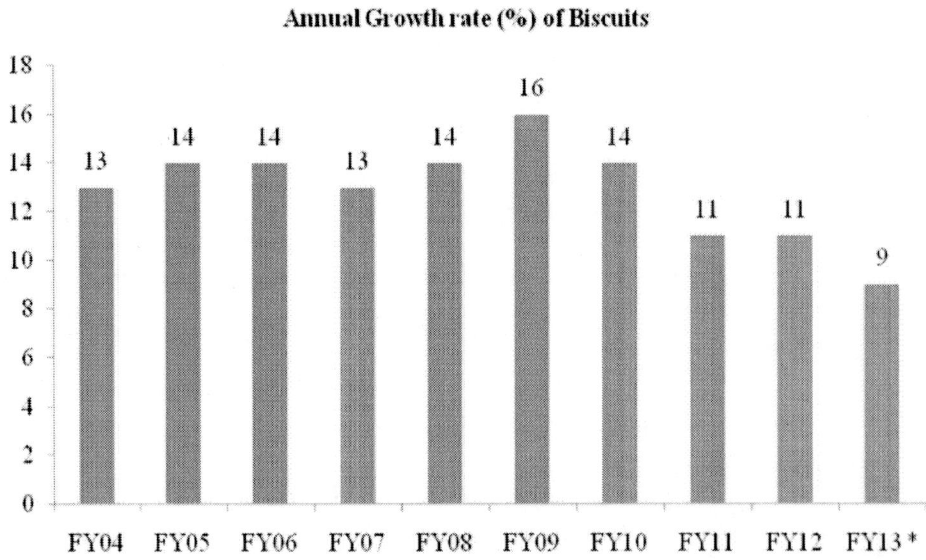

Note: * FY13 data as on 30 SEP 2012; FY represents financial year data.

Source: IBMA, 2013

Figure 1.1 Annual Growth rate of Indian Biscuits Industry

Source: IBMA, 2013

Figure 1.2 Key Brands of Biscuits Industry in India

India Biscuits Industry came into limelight and started gaining a sound status in the bakery industry in the later part of 20th century when the urbanized society called for readymade food products at a tenable cost. Biscuits were assumed as sick-man's diet in earlier days. Now, it has become one of the most loved fast food products for every age group. Biscuits are easy to carry, tasty to eat, cholesterol free and reasonable at cost. States that have the larger intake of biscuits are Maharashtra, West Bengal, Andhra Pradesh, Karnataka, and Uttar Pradesh. Maharashtra and West Bengal, the most industrially developed states, hold the maximum amount of consumption of biscuits.

Biscuit industry contribute more than Rs 8,000 crore to the fast moving consumer goods (FMCG) industry and provide a vast opportunity for growth, as the per capita consumption of biscuits is less than 2.1 kg in our country. India It is classified under two sectors: organized and unorganized. Branded/Organized to Unbranded/Un organized market share of biscuit has been 70 per cent for organized sector and 30 per cent for unorganized sector. Apart from Big 3 (Britannia, Parle, ITC) there are around 150 medium to small biscuit factory in India. Exports of Biscuit were 14 per cent of the annual production during the period of 2012-13. Imports of biscuits into India (mainly high end products) have not shown any significant growth during the last few years and have not affected production/sales by the Indian Biscuit industry. The Industry is now facing problem from increase of raw material price.

With Government value added tax (VAT) up to 12.5 per cent has added to their woes. Biscuit consumption per capita in India has grown to 2.1kg per capita in comparison to 10 kg per capita consumption in Unites States of America, United Kingdom and Europe. Biscuits are placed at a higher value chain than Bread in India. This is the reason biscuits occupy 60 per cent of the production capacity of bakeries. Cadbury, Nestle and Brooke Bond tried to enter the biscuit market but their efforts flopped. But things are changing now.

A battery of new players (Kraft, Pepsico, United Biscuits, Glaxo, and Danone) is waiting to enter the biscuit market. The domestic players are competing against each other in their respective product segments. Though the demand for glucose biscuits is strong, the market share continues to decline as healthier brands (fiber rich) are available. Surprisingly, the Indian market is still largely un-penetrated and this offers ample scope for growth. At present, it stands third in position among food industry in revenue generation.

Indian Retail Industry

Unlike urban markets, rural markets are difficult to predict and possess special characteristics. The featured population is predominantly illiterate, have low income, characterized by irregular income, lack of monthly income and flow of income fluctuating with the monsoon winds. Rural markets face the critical issues of Distribution, Understanding the rural consumer, communication and poor infrastructure. The marketer has to strengthen the distribution and pricing strategies. The rural consumer expects value for money and owing to has unsteady and meagre status of weekly income; increasing the household income and improving distribution are the viable strategies that have to be adapted to tap the immense potential of the market.

Rural markets face the critical issues of distribution, understanding the rural consumer, communication and poor infrastructure. The marketer has to strengthen the distribution and pricing strategies. it is uneconomical to access a large number of small villages with a very low population density spread over a large geographic area. Social norms, traditions, castes, and social customs have greater influence on the consumer behaviour in rural areas than in urban areas. Factors such as limited physical access, low density of shops, limited storage facilities, need for a large number of intermediaries in the distribution channel to reach the end customers, and low capacity of intermediaries to invest in business make the tasks of reaching rural consumers very complex.

Retailing is the world's largest private sector contributing to 8 per cent of the gross domestic product (GDP) and it employs one sixth of the labour force. The estimated retail trade is expected to be US $ 7 trillion (Sabura et al., 2012). Many countries have developed only due to retailing and presently there is a vast change in the retail industry. As far as India is concerned it contributes to 14 per cent of GDP and it is the second largest sector next to agriculture which provides employment to more number of persons. India is classified in to the fifth most attractive retail destination and second among the countries in Asia. Worldwide it is ranked as fifth most attractive retail destination. Retailing includes all activities involved in selling the products and services to the ultimate consumers. So this is said to be the last person in the channel of distribution.

Functions of Retailing

Retailing is supposed to provide:

- Product Utility
- Place utility
- Time Utility
- Ownership Utility

It provides final end products to the consumers, not raw materials, end products in usable form to the consumers. Thereby it creates product utility. Second one, it is given in the place where it is required by the consumer. That is, retailing outlets are open in the places according to the convenience of the customer and also based on the demand of the consumer. Third one, it creates time utility that is the shops are open as per the requirement of the consumer, whenever the consumers want to go and shop they can go and shop at a particular period convenient to them. next one, when the product is sold finally it creates ownership utility. So retailing is a marketing intermediary which creates product utility, place utility, time utility and ownership utility in providing goods and services to the consumers.

Classification of retail industry

Retail Sector can basically be classified in to two segments. One is organized segment and another one is unorganized segment. As far as India is concerned this organized segment contributes only to 3 per cent of the retail trade and the unorganized segment contributes to remaining 97 per cent of the retail trade. Because all these days people have been purchasing only from the street vendors and from the local shops and organized retailing was not in vogue in India. Only after 1991, after opening up of economy and due to liberalization, this organized sector has come to light and presently it exists in various formats like

- Super markets
- Hypermarkets
- Departmental stores
- Specialty stores

The distribution channel structure in India is largely traditional and quite unique. The major channel components are the retail network, wholesale network, and the logistics infrastructure. The retail network in India consists of over nine million outlets. These include traditional outlets like paan shops; grocers or kirana stores; general stores; specialised shops for footwear, clothing, jewellery, watches, mobile phones, and consumer durables; newer formats like supermarkets, hypermarkets, and online stores; and service outlets like fast food outlets, beauty parlours, fitness centres, coaching centres, and so on.

Traditional outlets are spread across urban and rural India but the newer formats are mostly located in urban areas. The penetration of organized retail in India which is less than 8 per cent is quite low even in comparison to other emerging markets. The average retail outlet in India is very small in terms of area, number of employees, and number of stock keeping units (SKUs) stocked. Traditional retail in India offers consumers a number of advantages like convenience, home delivery, credit, and personalized service. On the other hand, modern retail offers periodic promotional offers, lower prices, wider assortment, a better ambience, and higher quality brands. The continued existence of traditional retail in India has been based on factors such as lower rentals, lower labour costs, credit from suppliers, low or no liability on taxes, and a legal framework which prevented foreign direct investment (FDI) in retailing until quite recently. Traditional stores have managed to hold their own against organized Indian retailers so far by making some changes in their operating practices. It will be interesting to see how traditional Indian stores will fare in future after the entry of price aggressive international retailers.

The food processing industry s gaining lot market share n India. In the background of changing retail trends, understanding customers perceptions, building relationship and retaining customers has been identified as major source of competitive advantage. Businesses around the world are becoming increasingly competitive day by day. Therefore in order to generate more customers and retain existing ones, companies engage in various forms of activities which are known as "Relationship Marketing". One of the most important components of relationship marketing is "Customer Relationship Management". Payne and Frow (2005) described Customer Relationship Management (CRM) as, "a strategic approach that is concerned with creating improved shareholder value through the development of appropriate relationships with key customers and customer segments" (p. 168). By looking at the definition it can be said that CRM is about creating, maintaining, and sustaining relationships with customers.

One of the most commonly used CRM strategies of a company are loyalty programs, which are designed to give variety of benefits to increase customer loyalty. In present multi channel retailing environment, companies seek to provide satisfaction to customers through various loyalty programs. In many cases, retailers with very limited amount of products and services would form joint venture of loyalty program to encourage customers to visit the retail stores which are part of such loyalty programs. This demonstrates the importance of integrated CRM strategies within the organization in current business environment.

Relationship retailing

Building strong customer relationships has never been more essential to retailers as in the current scenario. Changes in the marketing environment comprise of fierce competition, emergence of e-commerce, complex multi-channel strategies, and a volatile consumer buying behavior. These factors have placed more emphasis than ever on the importance of customer loyalty. Dick and Basu (1994) describe customer loyalty as "biased repeat purchase behaviour" that is a key driver for nurturing a sustainable competitive advantage, especially during tough economic times.

Some interesting facts brought forward by Kotler and Keller (2008) towards customer retention reveal that acquiring new customers can cost up to five times more, than satisfying and retaining existing customers. The renowned 20-80 rule, states that eighty percent of a company earnings is produced by twenty percent of its core customers. Based on these findings, many of the most successful retailers have shifted their focus towards CRM strategies which largely embrace the use of loyalty schemes. The goal of such programs is to establish higher level of customer retention in profitable segments by providing increased satisfaction and value to certain customers (Egan, 2008).

According to Berman and Evans (2010), relationship retailing can be attained by establishing and maintaining long-term bonds with customers, rather than acting as if each sales transaction is a completely new encounter. It compels retailers to implement some specific strategies that not only differentiate their offer in ways that are relevant to their target market, but also difficult for competitors to duplicate (Berry and Gresham, 1986). Because consumers today recognize fewer real product differences and show less brand loyalty in their search for value, the following topic has being dedicated to further explore the benefits of a value-oriented retail strategy.

Building relationships by providing value

Kotler and Keller (2009) defined customer-perceived value as "the difference between the customer's evaluation of all the benefits and all the costs of an offering". It denotes a blend of quality, service, and price. From the retail perspective, value is typically described using the "value chain" concept, which encompasses the entire cluster of benefits – including all the activities, processes, tangible, and intangible attributes–offered to clients throughout the distribution channel (Berman and Evans, 2010). They range from logistics processes to the merchandising of goods, including store location, hours, atmosphere, sales personnel, and brands A value-oriented retailer can decide on the following strategies:

Expected retail strategy

Consists of the basic components expected by a certain consumer segment for the type of retailer such as product assortment, store cleanliness, and knowledgeable staff.

Augmented retail strategy

Comprises the additional features offered by a given retailer – like store loyalty programs and personalized services – with the aim of differentiating their offerings from direct competitors, and to enhance store patronage.

Potential retail strategy

This is characterized by the addition of exclusive value-added elements to a retail offer, providing a competitive edge to the firm in areas where other firms are not able to capitalize. According to Berry and Gresham (1986), some of the most powerful retail relationship programs feature two important foundations:

1. Relationship customization, related to personalized offer and individualized service, and
2. Offer augmentation, which entail differentiation through the enhancement of the retail offer. In the context of loyalty schemes, majority of retailers compete at the augmented level with similar merchandising and mimicked reward schemes. Competitive advantage is however held by those who go beyond customers' standard expectations and build on potential features.

It is important to mention that the ever changing digital revolution, driven by the spread of internet, mobile technology and recently the social networking phenomena, resulted in a new "Information Age" of consumers. Due to the ease of access to various information sources, which allows shoppers to compare a variety of available goods and services, this consumer segment is becoming very much price-and-quality sensitive, demanding "more for less" from the shopping experience. The effective use of relationship management strategies will allow retailers to consistently identify and communicate meaningful value chain elements to its market segment, which will ensure differentiation and build on loyalty.

CRM – The concept

CRM perhaps the most important concept in modern marketing until recently, CRM has been decide narrowly as customer data management activity .in its broader sense, CRM is the overall process of building and maintaining profitable customer relationship by delivering superior customer value and satisfaction. It deals with all the aspect of acquiring, keeping and growing customers. The key building lasting customer relationships is to create superior customer value and satisfaction.

Organization can build customer relationship at many levels depending on the nature of the target market. To build customer relationships, companies can add structural ties as well as financial social benefits. Today the companies are building more long and lasting relationship with more carefully selected customers, changing demographics, more sophisticated competitors mean that are fewer customer to go around. Many organizations are fighting for shares of flat or fading market .in a addition to being good at customer relationship management, marketers must be good at partner relationship, which mean working closely with goal departments outside the company to jointly bring greater value to customer. Successful CRM focuses on the profiling or understanding the needs and desire of customer and is achieved by placing these needs at the heart of the business by integrating them with the organization's strategy and goal.

Importance of CRM

The customer is the real king of market the one who makes the difference between a company's success and failure. CRM facilitates growth in that it aids the reduction of cost and increased availability of information to the organization that can be used according the customer requirements. The specific advantages derived through CRM are quoted below:

- Better customer satisfaction.
- Gaining competitive advantage.
- Enhancing customer satisfaction.
- Easy handling of daily task.
- Makes awareness of customer needs.

Therefore organizations are now capitalizing on this need and are realizing the value of CRM application.

Application of CRM

CRM application includes a wide range of sectors ranging from business, education, health. The presence of CRM shall be reflected in the following significant sectors:

- Healthcare
- Financial services
- Retail
- Real Estate
- Banking
- Insurance

Customer relationship is the essence of a successful business and that's why many firms started to adapt a CRM department which specializes in the sole purpose of sustaining its customer base in the market. Customer relationship is a major aspect for a long lasting business, especially in retailing as retaining the customer could assure a thriving business with the support of the customers. Retention strategy is not a luxury any more as competition is not getting any less. As "this strategy gives the firm the chance to have the merits of it as it is cheaper to maintain the existing customers than acquiring new ones and to enjoy a profitable association," (Reichheld and Sasser 1990), the result of this strategy is loyalty. Organizations need to understand the dimensions of this word and study techniques of how it is obtained, as many retailers link the repeated purchase with loyalty, whereas the difference in major as well as frequent purchase could be for any other reason such as convenience, lack of choice and information, or simply inertia.

CRM is not a system, or a technology, or a simple application. CRM is a holistic strategy that places customers at the centre of all business operations. Some call this customer-centricity; others coin this approach as listening to the "voice of the customer." Regardless of semantics, CRM is the integration of people, processes, and technology, which allow a company to leverage its information assets to identify, attract, and retain profitable customers. Effectively implementing a customer-focused business strategy requires a cultural shift that converts the focus from products to customer needs. Rather than heavily promoting products to create perceived needs, consider working collaboratively with customers to really understand how organizations can help them.

As companies strive to manage expanding product offerings, diminishing competitive differentiation, and increasingly powerful and demanding customers, the winners will be those products/brands that convince their customers to buy and keep more products and meet and exceed their customers' expectations at all critical customer touch points. Faced with this challenge, traditional product-oriented organizational structures and systems are simply no longer up to the job. An alignment of the customer relationship management (CRM) program and supporting organizational structures, processes, and technologies is required to achieve a sustained excellence in customer experience, service, and responsiveness.

As the organizations think about how the customers' experiences add up to create their overall brand experience, it's helpful to focus on the three most essential marketing objectives and the metrics that reveal how well an organization is meeting those objectives:

1. **Customer Acquisition:** with a goal of acquiring the right customers in a cost-effective way. Three critical customer experiences in the acquisition process are awareness, learning and persuasion.

2. **Customer Experience:** Marketers must focus on product "wow" in delivering a "wow" customer experience that exceeds expectations. Three critical customer experiences required for product "wow" are great first-time usage, usability and benefit delivery.

3. **Customer Retention:** Marketers must focus on customer retention—retaining and nurturing loyal customers, and turning them into advocates. Three critical customer experience elements

in the retention process are long-term usage and satisfaction, the purchase of more products and services, and positive word-of-mouth.

Evaluating these essential business-building drivers within the customer experience framework will help organizations focus on the most important levers for achieving marketing results.

Customer relationship management (CRM) is a system for managing a company's interactions with current and future customers. It often involves using technology to organize, automate and synchronize sales, marketing, customer service, and technical support. Customer relationship management entails all aspects of interaction that a company has with a customer, whether it is sales or service-related. While the phrase customer relationship management is most commonly used to describe a business-customer relationship, CRM systems are also used to manage business contacts, clients, contract wins and sales leads. Customer relationship management is often thought of as a business strategy that enables businesses to improve in a number of areas. The CRM strategy allows you to following:

- Understand the customer
- Retain customers through better customer experience
- Attract new customers
- Win new clients and contracts
- Increase profitably
- Decrease customer management costs.

The role of CRM in a challenging economy

In today's economy organizations can't rely on brute strength to maximize the value of their customer relationships; rather, they need enabling technologies and tools. CRM is a technology that allows organizations to track and leverage every customer interaction to maximize revenue opportunities and improve customer loyalty with respect to the retailer wholesalers and customers.

Five key strategies with CRM

While there are surely many approaches being espoused in the market today, we believe there are five main strategies that companies can employ to survive and thrive during uncertain economic conditions:

1. Focus on existing customers
2. Maximize revenue opportunities
3. Do more with less
4. Reduce operational costs
5. Optimize existing IT assets

Focus on existing customers

It is common knowledge in business circles that it is significantly cheaper to retain existing customers than to acquire new ones. By providing employees with quick access to actionable customer data, organizations can better identify the right customers, increase their loyalty, and maximize their profitability.

Maximize revenue opportunities

In today's economy it is imperative that organizations not only maximize the value of existing customers but also win new business in order to establish a foundation for sustainable growth. The effective ways to optimize the revenue opportunities are in terms of optimizing the marketing mix and re-prioritize sales investments.

Do More with Less

When money is tight, shrewd businesses look for ways to do more with less. And one of the fastest ways to achieve improved efficiency is by streamlining key business processes and improving individual productivity.

Reduce Operational Costs

Doing more with less is certainly valuable, but at the end of the day organizations need to reduce "hard costs". CRM can help with cost reduction in two primary ways. It facilitates delivery of cost-effective customer service, and it enables organizations to minimize IT costs through system consolidation.

Optimize Existing Assets

No one would dispute the importance of reducing hard costs in these times but another effective way to improve the bottom line is by better leveraging and extracting value from existing assets.

Thus, being one of the leading biscuit industry the analysis for Britannia will provides an vital inputs for the growth and development in near future this creates an interest in the evaluation of determine the sales that has been achieved by Britannia industries especially by the retailers in the Districts of Pathankot and Gurdaspur of Punjab in India. In addition to this the identification of various conflicts among the channel members and the techniques adopted by the distribution outlets also becomes vital for the future progress of organisations.

❑❑❑

Chapter 2

Review of Literature

Brand Coverage

Keller (1993) concluded that the main functions of a brand from the consumers' perspective is considered to be the minimization of perceived purchasing risk, which in turn helps cultivate a trust-based relationship.

Morgan and Hunt (1994) revealed that manufacturers will also win a reputable name through consumers' positive attitudes and strong brands are more committed to their retailers, and the retailers will in return invest more effort and resource in maintaining the relationship. Therefore, a strong brand leads to mutual trust and commitment, and fosters the manufacturer-retailer relationship.

Lassar *et al.* (1995) elaborated that brand equity facilitated the effectiveness of brand extensions and brand introductions. The study further presented a scale to measure customer based brand equity. The customer based brand equity scale is developed based on the five underlying dimensions of brand equity: performance, value, social image, trustworthiness and commitment. It has been concluded that the brands that scored higher on the customer based brand equity scale generally had higher prices.

Aaker and Joachimsthaler (2000) elaborated that brand identity is a unique set of brand associations implying a promise to customers and includes a core and extended identity. The study revealed that to be effective, a brand identity needs to resonate with customers, differentiate the brand from the competitors, and represent what the organization can and will do over time for better customer relationships.

Webster (2000) concluded that branding benefits to intermediaries such as retailers and wholesalers, a strong brand with high brand recognition and brand awareness speeds up the stock turnover rate, lowers the selling cost, and leads to higher sales.

Brucks *et al.* (2000) concluded that delivering quality products requires an understanding of the critical dimensions and cues that consumers use to judge quality. The study first developed a generalizable typology of quality dimensions for durable goods that includes ease of use, versatility, durability, serviceability, performance, and prestige. The study concluded that consumers use price and brand name differently to judge the quality dimensions, searching for price and brand name much more frequently when evaluating prestige than when evaluating any other quality dimension. Further the study suggested that managers must determine the relevant quality dimensions for a product category and the cues that are salient for judging those dimensions.

Ballester and Alemán (2001) focused on the roles of perceived quality, brand reputation and especially satisfaction, due to the fact that they summarized consumers' knowledge and experiences, guiding their subsequent actions. The study concluded that the shifting emphasis to relational marketing has devoted a lot of effort to analyze how other constructs such as trust predict future intention.

Kotler (2001) revealed in marketing literature great attention is focused on brand image which is studied from two: companies and consumer's perspectives. The contribution suggested that the approach of a company is directed towards the improvement of marketing activity, connected with strategies of positioning and retaining of a positive brand image. While on the other hand, consumer's approach is based on consumer's attitude towards the interpretation of brand image and brand equity. The study concluded that the importance of brand in the market is influenced by company's ability to evaluate the fact how consumers interpret the image of brands and company's ability to manage the strategy of brand positioning, adequately revealing brand's equity to a consumer.

Davis (2002) reiterated that consumers do not have a relationship with a product or service but he/she may have a relationship with a brand because a brand is a set of promises and therefore the strongest brands own a place in the consumer's mind.

Grundey (2002) suggested that identity should help in the development of the relationship between a brand and a consumer, providing a particular benefit. The study concluded that main principles of brand identity are communication that has been not an important supply of information but a daily struggle to remain heard and visible, which constitutes the prerequisite of durability; brand should adequately react to changes which makes up the prerequisite of understanding; brand should reflect consumer's benefit and expectations, remaining harmonious, which is regarded as the prerequisite of consistency.

Czellar (2003); Grime *et al.* (2002) stated that there has been a plethora of research on consumer attitudes to and evaluation of brand extensions and on their impact on the 'equity' of the parent brand.

Keller (2003) concluded that it is up to the consumers to determine a brand's strength, but in fact it is the branding process that creates a unique mental map in a consumers' mind and guides their behaviour towards repurchase.

Pelsmacker *et al.* (2004) revealed that the attributes of a branded product add value for consumers, the intermediaries, and the manufacturers. The study concluded that the most significant contribution of a strong brand to consumers would be the reduced searching time and cost when they are confronted with a set of identical products. It helped consumers to identify and locate a product with less information processing and decision time because of the expected quality from accumulated brand knowledge. The study additionally concluded that sometimes manufacturers may even gain greater bargaining power over their retailers, and are presented with more distribution channels.

Swait and Erdem (2004) described that a strong brand is a valuable asset to the company. A well recognized brand serves as a signal, and it increases the likelihood for consumers to place the product in their consideration or choice set.

Nandan (2005) suggests that strong brands have two very key distinct features namely brand image and brand identity however no matter how good a company is such as having a unique vision, strong management or superior product if the core benefits of the brand are not clearly communicated to the right target audience, the brand will ultimately fail.

Brand Comparison

Steiner (1993) emphasized that retailers use manufacturer brands to generate consumer interest, patronage and loyalty in a store. With the growth of competition, retailers compete with manufacturers for consumer pull to increase their relative market power and their share of the total channel profit.

Srivastava *et al.* (1998) stated that because of these loyal customers (i.e. implying higher customer retention rate), a company will find it easier to extend its brand, and lower its marketing costs.

Ehrenberg (1988) explained multi brand purchasing in terms of the similar functional outcomes of selecting among various competing brands. The study found that a more objective approach might be of more value to marketing and retailing managers. The study further revealed that unless marketing has an appropriate definition of substitutability and an operation measure, it cannot adequately address the brand level of analysis, that which makes marketing inquiry unique within the social sciences.

Davis (2000) reported that an understanding of competitors is vital in building a strong brand and the failure to understand one's competitors is ultimately the failure to know one's customers: who they are, how they think, and how the brand can be adapted to meet their needs.

Yoo *et al.* (2000) recognized that there are two types of marketing management efforts from a long term perspective of brand management namely: brand building activity and brand-harming activity. It was observed that frequent use of price promotions is a typical example of brand-harming activity whilst high advertising spending, high price and distribution through retailers with store images and high distribution intensity are good examples of brand-building activity. The results of regular price cutting can negatively affect brand equity as a perception is created that product quality has been compromised. It has further been suggested that managers should avoid frequent price cuts or a consistent low price strategy because they lower perceived quality and product image.

Ramanathan (2007) stated that accessing the market through the distribution network is important as it significantly increases the penetration of brands. The higher dealer penetration results in higher reach among consumers and therefore a higher market share. This is examined by comparing the percentage dealer penetration and market share.

Foxall, *et al.* (2010) elaborated in a comprehensive analysis to define brand substitutability and potential product clusters within the overall category. The results indicate that brands performed as expected as perfect substitutes for one another while making competitive analysis. Additionally, the results conclude that five subcategories of biscuits (chocolate biscuit count lines, chocolate coated biscuits, filled biscuits, plain sweet biscuits, and savory biscuits) generally performed as a separate product.

Bagchi and Khamrui (2012) attempted to examine the financial performance of two leading FMCG companies in India - Britannia Industries and Dabur India, over a period of ten years. The study revealed that large scale and low cost production, modern retailing strategies, branding and maintenance of intense distribution network have given FMCGs an edge over others in raising hovering revenues. The study further used various accounting ratios and statistical tools like, linear regression analysis and multiple correlation analysis. The results revealed that though Britannia Industries is passing through hard times in terms of profitability, Dabur India is enjoying its enhanced performance and continuous growth in the sector.

Khan *et al.* (2012) in the contribution compared five different brands of locally produced biscuits and five brands of foreign biscuits from the market. The study found that the moisture and water activity values have been increased with time for all the samples whereas sensory scores decreased. However, it has been analysed that all the foreign and local biscuit samples were found acceptable under the study period but only one local biscuit was found unsalable.

Customer Relationship Management

Frazier, *et al.* (1985) observed that manufacturer's support programs positively, though weakly to distribution intensity, when a manufacturer provides many support programs, retailers may have some incentive to join and remain in the channel system. Manufacturer coordination efforts are related inversely though weakly to distribution intensity, when close coordination is needed manufacturers are likely to limit distribution to reduce potential difficulties in channel operations and foster a supportive atmosphere in their exchange relationships. Retailer investment significantly weakens the inverse relationship between manufacturer coordination efforts and distribution intensity. Retailers making heavy investments are signalling to the manufacturers their willingness to sell and service it effectively.

Wileman and Jary (1997) in one of the major contribution identified the overwhelming success of some private labels in European retail (e.g. Marks and Spencer, Asda, Tesco, Carrefour) and the popularity of President's Choice private label in Canada. The study concluded that strong retail brands have become the source of intense interest and analysts found 50 per cent of all EU grocery sales were represented by the top ten retailers during 2005. Further, it has been revealed that the cumulative power carried by retailers and their store brands are significant (as corroborated by Lepir, 2001).

Dunner and Narasimhan (1999) revealed that private label strategy can help project a lower-price image for retailers, increase their bargaining power over manufacturers and producers of major national brands, and lead to increased control over shelf space.

Kotler (2000) assured that CRM uses IT to gather data, which can be used to develop information acquired to create a more personal interaction with the customer. In the long-term, it produces a method of continuous analysis and reinforcement in order to enhance customer's lifetime value with firms.

Baker and Ballington (2002) elaborated that country associations have also been positioned as a potential source of competitive advantage for manufacturers seeking to differentiate their product in domestic and/or export markets.

Chen and Popovich (2003) argued that CRM is a complicated application which mines customer data, which has been retrieved from all the touch points of the customer, which then creates and enable the organization to have complete view of the customers.

Coelho and Easingwood (2003) identified that the distribution tends to be one of the most immutable of the marketing mix decisions, but a number of external factors have led to an increase in its importance, namely, pressures on competitive advantage, the increased power of distributors, pressure to reduce distribution costs, a new stress on growth, and new technological developments.

Pracejus and Olsen (2004) showed that consumers are less sensitive to price and product attribute differences when a CRM campaign is used. The study identified that it is likely that consumer's over-report altruistic behaviour, and the real influence of CRM on purchasing decisions may in fact be lower. A study observing actual consumer behaviour is needed to confirm the findings of existing studies.

Gray (2005) in the contribution identified that Australia's largest food retailer in 2005 alarmed some New Zealand suppliers who were concerned that local products and brands would be replaced by house brands and Australian suppliers. It is further estimated from the results that only 20-30 per cent of grocery items sold in New Zealand are sourced offshore due to local tastes in food, the value of iconic national brands, as well as the cost efficiencies of local sourcing.

Wells (2007) in the competitive environment of supermarket retailing, packaging is becoming a significant weapon for FMCG firms and retail chains alike.

Mueller (2010) characterized customer relationship management aspect of the business as a highly dynamic, and convincingly argues that businesses have to adopt a proactive approach in devising relevant programs and initiatives in order to remain competitive in their industries.

Verhoef *et al.* (2010) depicted that many retailers have collected large amounts of customer data using, for example, loyalty programs. The study discussed how retailers can gather customer data and how they can analyze these data to gain useful customer insights. Further it has been revealed that the existing knowledge on the application of marketing actions in a CRM context should be taken into consideration while providing an in-depth discussion on CRM and firm value.

Tsai *et al.* (2010) revealed that if a business is to enhance its performance, it must have to realize importance of considering customer value, customer satisfaction and customer loyalty. The analysis further concluded that when the products have little difference in terms of features and quality and variety, then customer satisfaction and customer loyalty will play a vital role in increasing the rating of firm.

Peppers and Rogers (2011) suggested that there is global tendency in customer relationship management that relates to the shift from transactional model towards the relationship model. The study further argued that satisfying customer needs as a result of on-time transaction is not sufficient today in order to ensure the long-term growth of the businesses. The study additionally mentioned that businesses have to strive to maintain long-term relationships with their customers in order to maintain flexibility to adopt their increasing expectations and thus achieving their life-long loyalty. The study further stressed that, businesses that refuses to acknowledge this tendency in the global marketplace would be risking their market share and growth prospects in the future.

Cox's (2011) concluded that highly practical recommendations to retailers of various sizes in terms of increasing the levels of revenues through adopting a range of customer relationship management principles.

Obaji (2011) investigated that the distribution still offers a new frontier for competing successfully, especially if the emphasis is placed on the design and management of superior marketing channel systems to provide excellent customer services. Yet designing optimal marketing channel systems to boost sales, formulating innovative distribution strategies and managing channels system effectively is no simple task.

Sudha Vemaraju (2011) analysed that customers perceptions, CRM and changing waves in retailing are the three critical areas that companies should focus over the competitor analysis.

Hossein and Zakaria (2012) stated that CRM is a strategic approach that integrates people, business and technology to understand the needs of customers to be more satisfied. The study found eight benefits of CRM which are important and beneficial for customers such as improve customer services, increased personalized service, responsive to customer's needs, customer segmentation, improve customization of marketing, multichannel integration, time saving, and improve customer knowledge. The result helped the managers and companies to know and recognize the factors and benefits of CRM which are more significant for their customers. So they can increase their customer's satisfaction and retention by more attention to these benefits.

Kumar and Reinartz (2012) dealt with the aspects of CRM strategy implementation in relation to the management of different distribution channels through which companies interact with their customer. The study highlighted the challenges for implementation and maintenance of CRM with wholesaler and retailers. The study concluded that a clear instruction should be successfully provided by the management to the both multichannel systems and shoppers.

Sabura *et al.* (2012) in one of their major contribution examined retailer's attitudes towards Britannia products in the rural areas of Tirunelveli district. The study concluded that 36 per cent of respondents were highly satisfied, 27.2 per cent of respondents were satisfied and 16 per cent of respondents were neutral towards the dealers visit to the outlet. About 33.2 per cent of the respondents were highly satisfied, 32.4 per cent of respondents were satisfied with the audit visit to check the quality and expiry of the Britannia biscuits. 37.6 per cent of the respondents were highly satisfied and 43.2 per cent of the respondents were satisfied and none of the respondents were highly dissatisfied with the fulfilment of consumer's requirements. In addition to this, 35.6 per cent of the respondents were highly satisfied and 38.8 per cent were satisfied with the target assigned by the company, whereas, 29.2 per cent of the respondents were highly satisfied and 41.6 per cent were satisfied with the facilities provided to promote the sales. Moreover, majority of the respondents 94.4 per cent of respondents were satisfied and none of the respondents were highly dissatisfied about the price of the Britannia biscuits. Meanwhile it has also been concluded that 37.2 per cent of the respondents were highly satisfied and 48.8 per cent of the respondents were satisfied and none of the respondents are highly dissatisfied towards the offers and schemes provided by the company. On the other hand, majority of the respondents (55.2 per cent) were highly satisfied, 14.4 per cent of respondents were satisfied and none of the respondents are highly dissatisfied towards the replacement of the damages; 53.6 per cent of the respondents were highly satisfied, 35.2 per cent were satisfied and none of the respondents were dissatisfied about the availability of the Britannia products. Meanwhile in terms of commission paid by company 38 per cent of the respondents were highly satisfied and 48.4 per cent of the respondents were satisfied and none of the respondents were highly dissatisfied towards the commission paid by the company. Finally the study experienced that more than 60.0 per cent of the respondents were satisfied with the credit policy, service quality and supply of products to the retailers. Additionally, the analysis of variance inferred that the three variables, customer relationship management, marketing mix and supply chain have an equal effect in determining the retailer's attitude towards Britannia biscuits.

Lo (2012) CRM activates help in building long lasting relationships and these relationships give company joy of retained customers, retailers and wholesalers. Relationship marketing is a way to obtain trust and satisfaction, which in turn ensures sustainable success of an organization.

Jumaev *et al.* (2012) confirmed that loyalty is the most desirable outcome for any firm, as it ensures continuous inflows of customers and ultimately of profits. The analysis of the data showed that all the predictors have positive correlation to customer loyalty.

Alshurideh, *et al.* (2012) found that experience has a direct impact on future purchase behaviour. Positive experience is directly related to repeat purchase behaviour, and negative experience is directly related to switching.

Moolla and Bisschoff (2012) evaluated the model to measure brand loyalty. The results obtained from the research in this article showed that the model to measure brand loyalty were successful and remained valid in the case of all three the products. The results are promising, showing that the model to measure brand loyalty strives towards a generic tool that can be applied in the FMCG industry in different products. In addition, the model sets a scientific base for academia who wants to further research the concepts of brand loyalty and brand management.

Dancer (2013) found that distributors are aware of CRM and many have already adopted CRM tools. It has been experienced that the use of CRM ranges from simple contact and lead management tools all the way through robust pipeline, forecasting, and social networking or communication applications. CRM has traction with wholesaler-distributors because it fills a gap – it gives them the ability to manage sales by using data and metrics. The study further revealed that more and more distributors are using CRM because they have discovered CRM can help them improve their profitability.

Ejaz *et al.* (2013) tried to understand that how changes in the level of CRM practices and their effectiveness influence customer satisfaction and loyalty. Results indicated that CRM has only indirect impact on customer loyalty and word of mouth.

Nagasimha Kanagal revealed that competitive marketing strategy (CMS) has relationship marketing (RM) as one of the key functionality in enhancing business performance. The role of RM in CMS includes:

guide moments of truth, improve profitability, build partnering, address 'Customer Better', buy in of customer attention, protect emotional well being, understand consumer psyche, build trust with customer. All these roles are observed empirically in the hotel industry, with some hotels placing emphasis on their extraordinary operations and services to engage with the customer.

Importance of the Study

To determine the retailers' attitude toward the customer relationship management so as to develop appropprtaite policies, the present endeavour plays a vital role. The present study has provided the insight about Britannia Industries performance in biscuit sector in different beats of Pathankot and Gurdaspur districts of Punjab region. The study has given in depth analysis about Britannia industries in biscuit sector as far as their availability and non availability of different variants of biscuits, infrastructure, their inventory modelling and their warehouses are concern. The present contribution will additionally help the distributors to make their performance up to the blot and thereby make our the appropriate strategies in order to enhance the market share for the Britannia biscuits in terms of different variants compared to the competitors in the rural areas of Punjab.

One of the important aspects of this study is to increase the market segment for the product. The study has attempted to provide the strategy makers of Britannia industry with some sales and marketing inputs for improving their customer relationship management and market share with respect to Pathankot and Gurdaspur districts of Punjab. Moreover, CRM strategies will facilitates growth that aids in the reduction of cost and increased availability of information to the organization that can be used according the customer requirements.

Objectives of the Study

- To find out Britannia Coverage in Pathankot and Gurdaspur districts of Punjab.
- To compare and check the Brand availability of the Britannia Biscuits at different stores.
- To identify the impacts of CRM on the retailer's attitude.

Scope of the Study

The present study is limited within the retail outlet making operations for Britannia and other competitive products in Punjab. The study has been confined to the 5 beats each from Pathankot and Gurdaspur districts respectively. Being the vital, most developing and potential rural areas for Biscuits of Punjab, the study tried to identify the attitude of retailers in both the districts. The study additionally elaborated the availability and non availability of the Britannia and the other variants of competitors from 175 retail outlets in Pathankot and 125 retail outlets in Gurdaspur. Although the focus of this study is on the identification of retailers' attitude towards the customer relationship strategies adopted by the distributors of Britannia, it is important to see availability and attitude of retailers towards the ongoing distribution aspect of an organization's, thereby providing the suggestions to serve by making coverage of untapped outlets.

Limitations of the Study

1. The study relates only to the retailers of selected areas of Pathankot and Gurdaspur districts of Punjab therefore, the findings cannot be true representation of all retailers (total population).
2. The personal bias of respondent has also been a limitation.
3. Respondent's ignorance to certain questions also posed as barrier towards certain responses.
4. Since the researcher has used structured schedule, the respondent's responses were immediate and the time taken for deciding was less which led to responses with not much thought. Therefore time constraint has also been a limitation in this study.
5. Illiteracy of the respondent posed a barrier for making them understand the question. Therefore, the responses cannot be taken factual.

Methodology

The main goal of this study is to identify cause and effect relationship between dependent and independent variables. The sampling structure and techniques adopted during the course of investigation have been described in this chapter.

Locale of Study

To analyze the status of availability and non availability of Britannia biscuit viz on viz their competitors, the present study was conductor in the rural area of Punjab region. As it is not compatible to cover the whole rural segment of Punjab, thus being the most efficient areas for the sale of Britannia biscuits, the study had been confined to the Pathankot and gurdaspur districts of Punjab. As rural population will depict the real scenario of Britannia in Punjab, thus for the benefit of policy makers it becomes viable to evaluate the scenario of the biscuits of Britannia in these regions. The information has been collected from the retailers running operations in various geographical areas of Pathankot and Gurdaspur districts of Punjab. The respondents were queried about the strategies adopted by the company to increase the sale at retail outlets of Punjab.

Type of Study

Although there are various tests of research studies mentioned in literature for conducting study in different disciplines. This study has been based on purposive type of research. A research design provides the frame work to be used as the guide in analyzing the data experiences with different research design will generally provide the capability to match a research problem with an appropriate design.

To collect the relevant information among the respondents in different beats of Pathankot and Gurdaspur districts of Punjab, the present study make use of a purposive type of research design. This is the most commonly used research design in the marketing research. This is one short research study at a given point of time and consist sample of the population of interest as the present study has the special emphasis on the respondents chosen from a defined target population that is 5 beats from each district i.e Pathankot and Gurdaspur of Punjab region for the study.

Therefore to collect the data from the retail outlets on the products of Britannia for analyzing the effectiveness of distributors, such type of research design seems to be appropriate. In addition to this a causual research design is also used to establish causual as for as possible by employing control and conditions under which the study can state with reasonable confidence whether or not dependent variables are affected by independent variables.

Sampling Technique

The present study has adopted a cluster random sampling to measure the influence of various CRM strategies on retailer's attitude. A cluster is a group of sampling units or elements which can be identified, listed, a sample of which can be chosen. Theoretically, a cluster could be on the basis of any criterion. But in practice cluster tends to found either in terms of geographical areas or membership of some groups. As the present study has chose the cluster on the basis of geographical area, thus, can also be termed as area sampling. The present study used a multi stage procedure to form the clusters. At first stage all available clusters were prepared and finally two were selected on the basis of extent of rural population. The first cluster was Pathankot and second was Gurdaspur. At second stage, five beats from each cluster were chosen on the basis of decision of researcher. Finally, 35 and 25 sampling units were selected in the cluster to be a part of a final sample.

Sample Size

The sample used in this study will consist of the retailers having retail outlets in different areas of Pathankot and Gurdaspur districts of Punjab. The data will be collected through filling up the semi structured schedule from retailers through market survey. The sample size for the study will be taken as 300. As the present study used the sample consists of 10 beats showing the status of availability of the Britannia biscuits in the rural areas of Punjab. To collect the information among the respondents that is retail outlets in 5 beats from each district or cluster a parameter that was taken into consideration was that district Pathankot comprises the sample size of 35 retail outlets from each beat and district Gurdaspur comprises the sample size of 25 retail outlets. Therefore, 300 sample size seems to be enough for conducting present study.

Districts	Beat	No. of Retailers to be covered	Total = (Beat*No. of Retailers to be covered)
Pathankot	5	35 per beat	5*35 = 175
Gurdaspur	5	25 per beat	5*25 = 125
Total	10		300

Note: - The beat represents the areas where distributor visits once in a weak.

Data Collection

The data collection is to collect important information to keep on record for further use, to make important decisions about different issues, and to pass vital information to others. The present study has adopted both the primary as well as secondary data collection technique. The research was conducted with the help of semi structured schedule that incorporates information related to the availability of biscuits and the impact of various strategies of customer relationship management

- **Secondary Data**: The data which has already been collected, complied and presented earlier by any agency has been used for purpose of investigation. The data will be collected through websites, journals related to brand and customer relationship management, newspaper articles related to marketing, magazines of finance and market and annual reports published by Britannia at regular interval of time.
- **Primary Data**: The first hand information has been collected through schedule. The study has incorporated a semi structured schedule that had taken into consideration various variables for achieving the basic objectives of the present study.

Variables involved in the schedule

For objective 1: To find the coverage of Britannia, the data has been collected retailer wise, beat wise and district wise. To collect the relevant information the retailers were supposed to answer the questions based on the availability and non availability of the Britannia biscuits in their respective stores. The revenant information thus collected were coded on the dichotomous scale with "1" showing the availability and "0" predicts the non availability of Britannia biscuits.

For objective 2: To compare the brand availability and different variants of Britannia biscuits, the data has been collected brand wise, retailer wise, beat wise and district wise. Following are the variables that will be used to identify the brand availability of Britannia with comparison to others.

- Name of the shop owner.
- Whether the retailer keeping Biscuit of any company.

- If not keeping whether they are interested to keep
- How much Biscuits brand the retailers have
- Space share by Biscuits
- Glucose Biscuit Depth
- Salted Biscuit Depth
- Cookies Biscuit Depth
- Marie Biscuit Depth
- Cream Biscuit Depth

The relevant information thus collected has been compiled on the basis of nominal scale and the information reveals the clear interpretation for the brand availability of different biscuits with respect to the other competitors.

For objective 3: To identify the impacts of customer relationship management on the retailer's attitude, following variables will be considered:

- Customer relationship management
- Retailers attitude
- Focus on existing customers
- Optimum utilization of resources
- Reduce operational cost
- Optimize existing assets
- Maximize revenue opportunities

To measure influence of **customer relationship management** following are statements measured on five points Likert scale:

1. I have tried the brand and would use again but have not done so in the past 3 months.
2. I have used the brand in the past 3 months but am not a regular user.
3. I am a regular user but this is not my most often brand.
4. I use this brand most often even though I do use other brands.
5. I always use this brand as long as it is available.

These statements were considered finally as the determinant for dependent variable at the stage first and then as the independent variables at the stage 2. Further, to measure **Retailers attitude** following are statements measured on five points Likert scale:

1. I feel that Britannia is a good company.
2. I have favourable attitude towards company.
3. I have positive attitude towards Britannia.

To determine the influence of independent variables namely focus on existing customers, optimum utilization of resources, reduce operational cost, optimize existing assets, maximize revenue opportunities following are the statement consider for analysing the response of the respondents at different location of Punjab.

Focus on existing customers

1. Distributors services
2. Britannia is aware of the problems related to the retailers.
3. Britannia is making the continuous or periodic visit to the retail outlets.
4. Britannia is retaining the profitable customers only.

Optimum utilization of resources

1. Streamlining of business processes should be there.
2. RTM should be launched.
3. Britannia is providing the actual products demanded by the retailer.
4. Britannia is taking much time in supplying the order placed by the retailer.

Reduce operational cost

1. Britannia has higher operational cost compared to other brands.
2. Britannia needs to reduce the operational cost.
3. Automation of services is beneficial.
4. Empowerment of distributor can be helpful in reducing the time and cost.

Optimize existing assets

1. Britannia replaces the expiry products more quickly than their competitors.
2. Britannia covers the breakage of products than other brands.
3. Sales distributor should be accessed with inventory data on daily basis.
4. Britannia is providing incentives to those retailers who achieve their targets.

Maximize revenue opportunities

1. Britannia is promoting its brands properly than others.
2. Britannia is providing sound or better schemes and programs to the retailers.
3. Britannia is properly implementing the Lakshya program.
4. KAT program is beneficial for retailers.

The information hence collected from the respondents using these statements was coded on the scale from 1 to 5, where 1 represents strongly dissatisfied, 2 represents dissatisfied, 3 represents neutral, 4 represents satisfied and 5 represents highly satisfied. In addition to this the negative statements were coded in the reverse manner. The study makes use of SPSS 14.0 to have the clear analysis for the policy makers of the organisation.

Methods of Analyzing Data

After collecting the data the results has been analyzed by using descriptive statistics, factor analysis and regression analysis technique. Descriptive statistics mainly mean, median, standard deviation, coefficient of variance, max, min. and factor Analysis will be used to minimize large number of variables into meaningful and correlated form.

Arithmetic mean

The arithmetic mean is calculated by adding a group of numbers and then dividing by the count of those numbers :

Mean = $\Sigma (X_1, X_2 \ldots X_n)/N$

Where

X_1, X_2 ....X_n are the observation on coverage of Britannia in different districts of Punjab retailer wise and beat wise.

N= number of respondents

Median

Median which is the middle number of a group of numbers; that is, half the numbers have values that are greater than the median, and half the numbers have values that are less than the median.

$$\textbf{Median} = \frac{\sum (N+1)}{2} \textbf{ th item}$$

Standard Deviation

The standard deviation is a measure of how widely values are dispersed from the average value (the mean).

$$\textbf{Standard Deviation } (\sigma) = \sqrt{\frac{\Sigma(x-Mean)^2}{(n-1)}}$$

Where n is the sample size.

Coefficient of variation

Coefficient of variation depicts the variation of the value in terms of percentage figure from the sample mean

Coefficient of Variation = σ/Mean*100

In addition to these descriptive statistics the present study will also incorporate the chart representation for the coverage of Britannia brand for different retail outlets surveyed in various areas of district Pathankot and district Gurdaspur of Punjab. The analysis will highlight the market share and brand availability of Britannia products retailer wise, brand wise, beat wise for Britannia industries ltd.

Factor analysis

Factor analysis is a very useful method of reducing data complexity by reducing the number of variables being studied factor confirmatory test has been used to minimize large number of variables into meaningful and correlated form. There are two stages in factor analysis.

Stage 1 can be called factor extraction process. Where the objective is to identify how many factors will be extracted from the data. The most important popular method used in the present study is principle component analysis. On the basis of computation of Eigen values. The number of factors extracted has been determined. The Eigen values have been translated approximately to the amount of variance explained by the factor.

The stage 2 is called rotation of principle component matrix. This is done for identifying which factors are associated with the original variables. The factor matrix provides us the loading values between 0 and 1. Values close to 1 represent high loading value and value close to 0 represent low loading. In the present contribution the variables with loading weight of 0.45 have been taken into consideration.

$$F_i = w_{i1}X_1 + w_{i2}X_2 + w_{i3}X_3 + w_{ik}X_k$$

Where

F_i = Estimate of ith factor

w_i = weight /factor score coefficient.

X_i= standardized variable

K = number of variables

Correlation and Regression

In addition to this the correlation technique and Regression analysis will be adopted for the comparative analysis of Britannia with others. Correlation depicts the relationship between two or more than two variables in a particular data set. It is represented by "r" and the value lies between -1 to +1. It also measures the strength of association between two variables. There are various methods for calculation the correlation coefficient. The present study will adopt Karl Pearson's Coefficient of Correlation Method.

$$r = \frac{\Sigma(x-\bar{x})(y-\bar{y})}{\sqrt{\Sigma(x-\bar{x})}\sqrt{\Sigma(y-\bar{y})}}$$

Where $(x-\bar{x})$ and $(y-\bar{y})$ is deviation of x and y items from their mean values

Regression takes a group of random variables X's (independent variables), thought to be predicting Y (dependent variable) and tries to find the mathematical relationship between them. Multiple regression analysis technique will be adopted in the present contribution to establish the relationship among the variables. In addition to this the regression coefficient is denoted by (R^2) that measures the association indicating the percent overlap between the predictor and the criterion variable.

Model to be Adopted at stage 1

Y1 = f (X1, X2, X3, X4, X5) – – – (1)

Y= a + β_1 (FOEC) + β_2 (OUR) + β_3 (ROC) + β_4 (OEA) + β_5 (MRO)

Where

Y = CRM

a = constant

β_1, β_2, β_3, β_4, β_5 = slope or coefficient of FOEC, OUR, ROC, OEA and MRO

FOEC = Focus on existing customers,

OUR = Optimum utilization of resources,

ROC = Reduce operational cost,

OEA = Optimize existing assets

MRO = Maximize revenue opportunities

Here, CRM is dependent variable and others are independent variables. Additionally.

FOEC = f (Distributors services, Britannia is aware of the problems related to the retailers, Britannia is making the continuous or periodic visit to the retail outlets and Britannia is retaining the profitable customers only).

OUR = f (Streamlining of business processes should be there, RTM should be launched, Britannia is providing the actual products demanded by the retailer and Britannia is taking much time in supplying the order placed by the retailer).

ROC = f (Britannia has higher operational cost compared to other brands, Britannia needs to reduce the operational cost, Automation of services is beneficial and Empowerment of distributor can be helpful in reducing the time and cost).

OEA = f (Britannia replaces the expiry products more quickly than their competitors, Britannia covers the breakage of products than other brands, Sales distributor should be accessed with inventory data on daily basis and Britannia is providing incentives to those retailers who achieve their targets).

MRO = f (Britannia is promoting its brands properly than others, Britannia is providing sound or better schemes and programs to the retailers, Britannia is properly implementing the Lakshya program and KAT program is beneficial for retailers).

Model to be Adopted at stage 2

$$Y = a + \beta_1 (CRM) \quad --- \quad (2)$$

Where

Y = Retailers attitude

a = constant

β_1= slope or coefficient of CRM

CRM = Customer relationship management

Where retailer's attitude is dependent variable and CRM is independent variable.

Retailers attitude = f (I feel that Britannia is a good company, I have favorable attitude towards company and I have positive attitude towards Britannia).

CRM = f (I have tried the brand and would use again but have not done so in the past 3 months, I have used the brand in the past 3 months but am not a regular user, I am a regular user but this is not my most often brand, I use this brand most often even though I do use other brands and I always use this brand as long as it is available).

Results

To find out the coverage of Britannia in Pathankot and Gurdaspur Districts of Punjab, the beat wise analysis has been conducted in the present study. 5 beats have been taken out from both the Districts to analyse the coverage of Britannia biscuits in Punjab. **Figure 1** depicts the coverage of Britannia in 5 different beats of Pathankot namely Dangu road, Dalhousie road, Dhaki road, Lamini and Mission road and Railway road. It has been analysed that each beat comprises of sample size of 35 retail outlets hence covering 175 retail outlets in Pathankot region. The study also revealed that in Dangu road beat out of 35 the Britannia biscuits were available at 32 retail stores and not available at 3 retail stores. Similarly, in Dalhousie road beat 31 retail stores were having the Britannia and 4 retail stores were not having it. However the Dhaki road beat shows the presence of Britannia in 32 retail outlets and absence in 4. While moving towards Lamini and Mission road the availability increases to 33 and not availability to 2. At last, in Railway road the availability of Britannia biscuits was in 31 retail stores and not available in 4. In nutshell, it can be concluded that the Lamini and Mission road shows the major coverage for Britannia in Pathankot district of Punjab. Meanwhile, it is suggested that, the Dalhousie road and Railway road should be focused more by the marketing personnels so as to enhance the coverage of Britannia in Pathankot District of Punjab.

Furthermore, the positive sign for Britannia in Pathankot District is the number of retail outlets in different beats. In totality there are 159 retail outlets that are having the availability of Britannia biscuits whereas rest 16 outlets show the unavailability. Therefore, the marketing and sales team should make the appropriate strategies so as to cover the rest 16 outlets and thereby make 100 per cent market coverage leading towards the increase in the sale of Britannia.

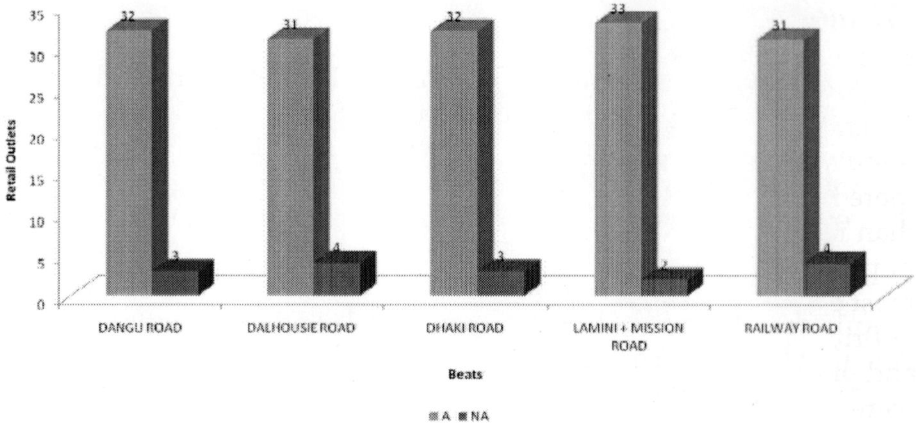

Source: Primary data.

Figure 1. Britannia coverage in District Pathankot

Note: **A** represents Availability and **NA** represents Non-Availability

Figure 2 depicts the coverage of Britannia biscuits in 5 different beats namely Sangalpur road, College road, Jail road, Babri Mani and Geeta bhawan road. From the **Figure 2**, it has been analysed that each beat comprises of sample size of 25 retail outlets. In Sangalpur road beat out of 25 retail outlets the Britannia biscuits were available at 22 retail stores and not available at 3 retail stores. In College road beat 19 retail stores were having the Britannia and 6 retail stores were not having it. However, the Jail road beat shows the presence of Britannia in 21 retail outlets and absence in 4. Similarly approaching towards the Babri Mani beat the availability of Britannia biscuits were in 20 retail outlets and not availability in 5. At last in Geeta bhawan road, the availability was in 21 retail stores and not available in 4 retail outlets. In nutshell, it can be concluded that the Sangalpur road shows the major coverage for Britannia in Gurdaspur district of Punjab. On the other hand the other beats especially College road should be focused by the marketing personnel so as to enhance the coverage of Britannia in Gurdaspur district.

Furthermore, the study reveals that there are 103 retail outlets that are having the availability of Britannia biscuits whereas rest 22 outlets show the unavailability. Therefore, the marketing and sales team should make the appropriate strategies so as to cover the rest 22 outlets and thereby make 100 per cent market coverage leading towards the increase in the sale of Britannia in Gurdaspur district.

In addition to this, it has been analysed from the statistics that Britannia is having more coverage in Pathankot district as compared to Gurdaspur district. However the probable reason for the less coverage in Gurdaspur district is due to the presence of more rural population then in Pathankot. Moreover the analysis conclude that there is a less number of retail outlets available in Gurdaspur district than in Pathankot leading towards decrease in market coverage and in particularly to the sales of Britannia biscuits. However, one of the most essential reasons for the less coverage and more non availability of Britannia biscuits in Gurdaspur district is due to the strong holding and presence of local brands especially Cremica. As these local brands were having less cost and more margin for the retailers in different beats of Gurdaspur district than in Pathankot. This has lead towards the presence of more local brands availability for biscuits than Britannia in Gurdaspur.

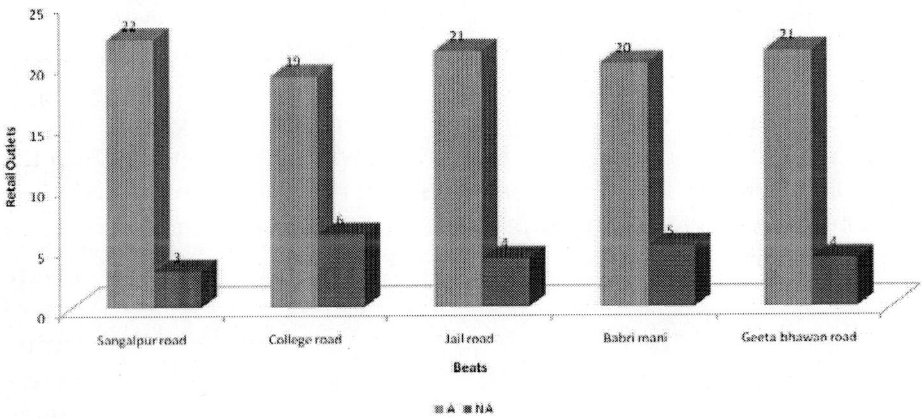

Source: Primary data

Figure 2. Britannia coverage in District Gurdaspur

Note: **A** represents Availability and **NA** represents Non-Availability

To confirm the presence of competitors with respect to Britannia, the present study also made check for the relative comparison of brand availability of Britannia with others. To analyze these results the respondents (retailers) have been asked to provide the response in terms of the availability of Britannia products/biscuits at respective outlet viz on viz competitors. From the **Figure 3**, it has been concluded that in district Pathankot Lamini and Mission road is depicting the presence of competition among the local brands also. The analysis revealed that Britannia is having the presence of availability of products at 33 retail outlets followed by Cremica (31), Parle (30), ITC (25) and others (23) respectively. However, it has also been analysed that Britannia is giving tough competition to its competitors in all the beats except Dangu road where the position of Britannia (32) to its major competitor Cremica (33) appears to be on second position. It's further been concluded that there is very high competition among the availability of Parle (31) and Britannia (32) biscuits for the retail outlets in Dhaki road.

On the other hand, it has been analysed that Britannia as a market leader for biscuit industry can focus on all retail outlets in different beats as the non–availability of Britannia products compared to other national and local brands is comparatively less except in Dangu road (3). It has been analysed that the marketing and sales team should focus more on Dangu road where the relative share of non- availability of Cremica (2) is less than Britannia.

In nutshell it has been suggested that the Britannia should make the appropriate strategies and analyse their potential compared to their competitors so as to deliver the timely and more marginal product to the retail outlets in different beats. This will help in building up the better relationship among the retailers and the sales team of Britannia thereby lead towards the improvement in the sales and revenue of both partners.

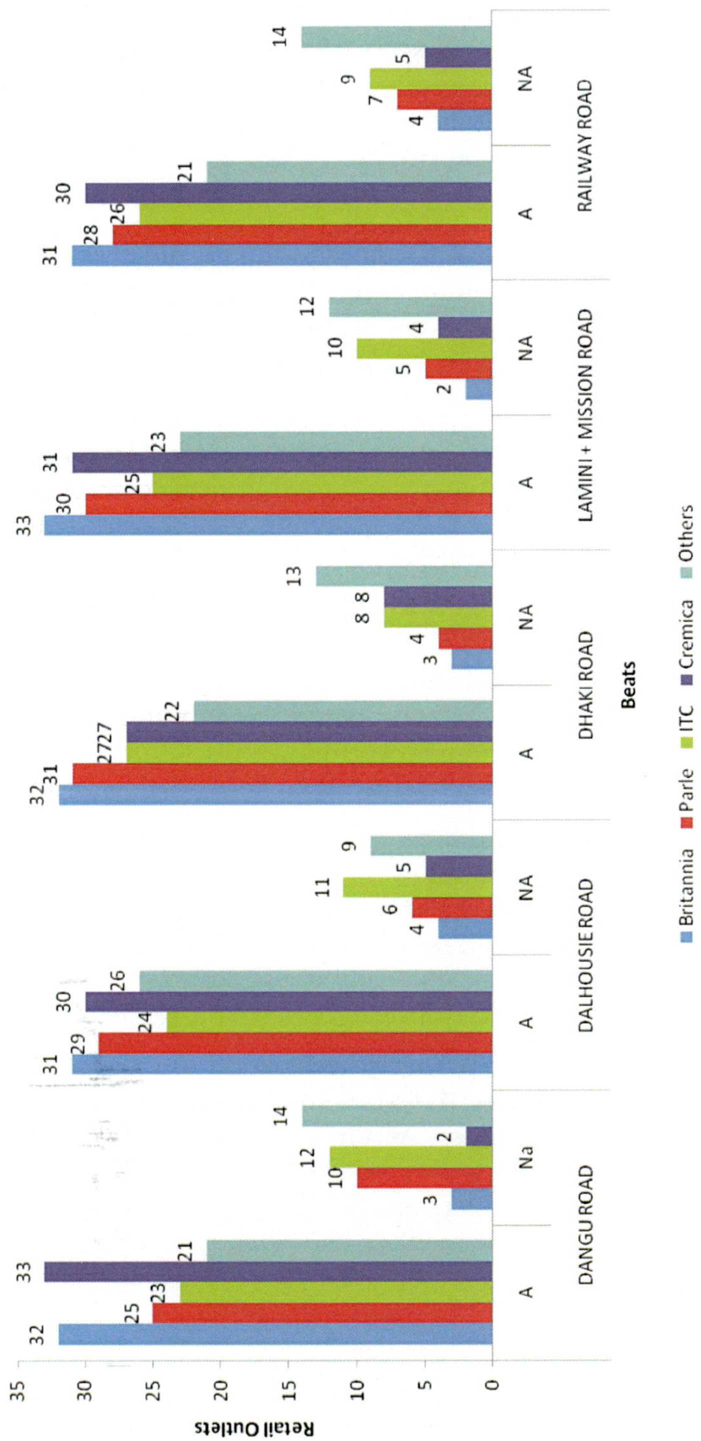

Figure 3. Brand availability of Britannia and its competitors in Pathankot district

From the **Figure 4**, it can be concluded that in district Gurdaspur where the availability and the presence of Britannia was less than Pathankot district have noticed that College road (19) and Babri Mani (20) are indicating the less coverage and low variety of Britannia biscuits as compared to competitors. It has been analysed that Britannia is giving tough competition to its competitors in all the beats except College road where the position of Britannia (19) is not good to its major competitor Cremica (20) and other local brands (21). It's further been concluded that there is very high competition among Britannia and local brands especially Cremica for the retail outlets in all the beats except college road where the local brands and Cremica are leading the market.

On the other hand, it has been analysed that Britannia as a market leader for biscuit industry can focus on all retail outlets in different beats as the non-availability of Britannia products compared to other national and local brands is comparatively less except in College road. It has been analysed that the marketing and sales team should focus more on College road.

In addition to this the Britannia can make the appropriate strategies and analyse their potential compared to their competitors so as to deliver the timely and more marginal product to the retail outlets in different beats especially the beat of College road.

Consequently Britannia should focus more on Lamini and Mission road followed by Dhaki road because of more availability of biscuits in the regions. They can further reduce the margin of non-availability to nil if the appropriate policies will be adopted. These policies can be in terms of advertisements, promotional activities, increasing margin for retailers and increase the visits of sales personnel to the retail outlets.

So it has been suggested that the Britannia should make the appropriate strategies and analyse their potential compared to their competitors so as to deliver the timely and more marginal product to the retail outlets in different beats. This will help in building up the better relationship among the retailers and the sales team of Britannia thereby lead towards the improvement in the sales and revenue of both partners in both the districts.

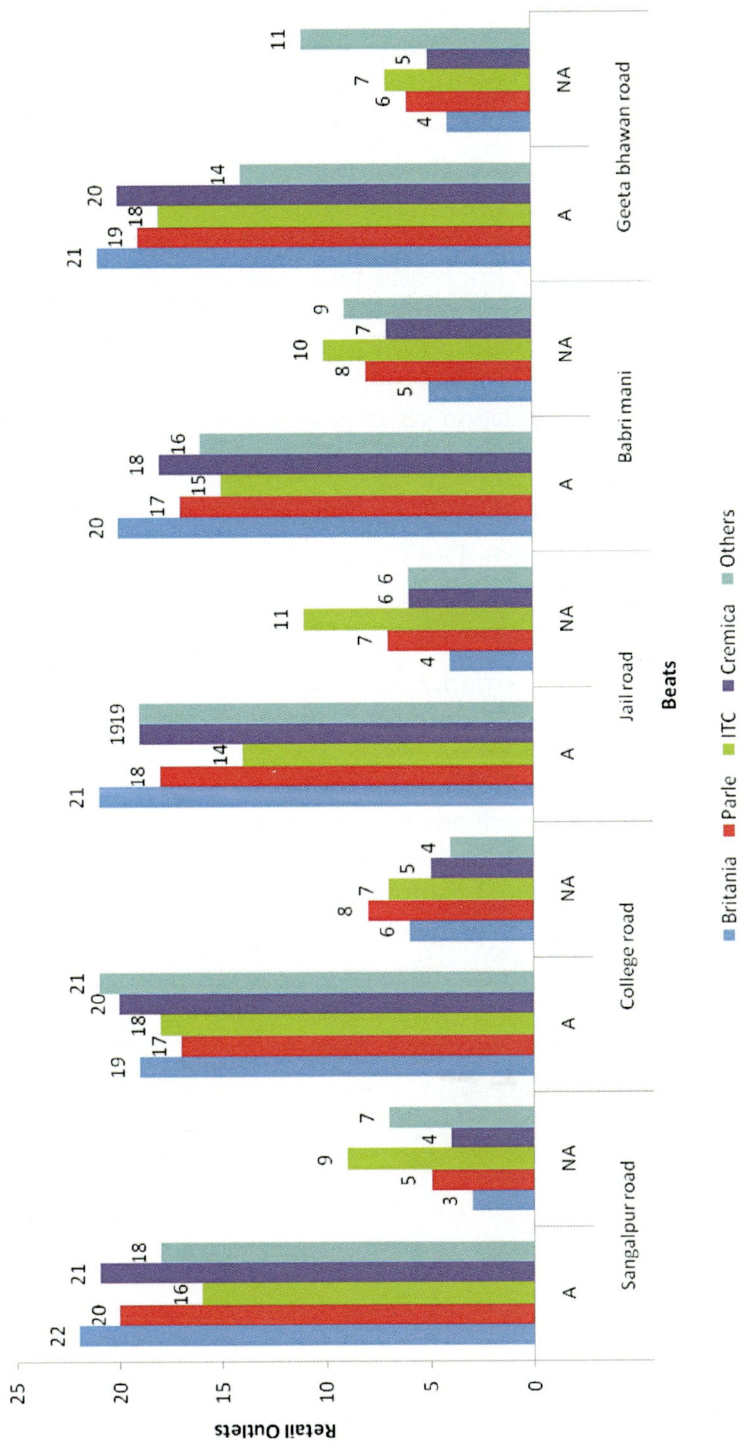

Figure 4. Brand availability of Britannia and its competitors in Gurdaspur district

To have a more elaborative and comprehensive picture of Britannia biscuits in different beats in the respective districts a relative comparison for the availability and non-availability of different variants have been analysed.

From **Figure 5**, it has been analysed that the variety of Glucose biscuits is leading the other variants in all the beats of district Pathankot except the beat of Railway road where the availability of Cream biscuits is more than other variants. Glucose is followed by GoodDay and 50-50 biscuits in all the beats except the Railway road beat where the variant of Cream biscuits is leading all other variants of Britannia biscuits. The availability of Cream biscuits is quite good in the beats of Dalhousie road, Lamini & Mission road and Railway road but not good in the beats of Dangu road and Dhaki road. So the company needs to focus on these two beats in order to make Cream biscuits more available in the market. Milk biscuits of Britannia are also not available that much as compared to Glucose and GoodDay so the company also needs to focus on this variant in order to make it more available.

There are only two such variants of Britannia biscuits which are found highly not available in all the beats of district Pathankot. These two variants are Marie and NutriChoice, these variants are more unavailable in the market due to their high price as compared to other variants. Thus the company needs to consider these two variants very much. because the main reason for the non-availability of these two brands is their size and price. So Britannia can increase the availability of these variants by giving more margins to retailers and also by introducing small size packs of these variants in order to increase the sales.

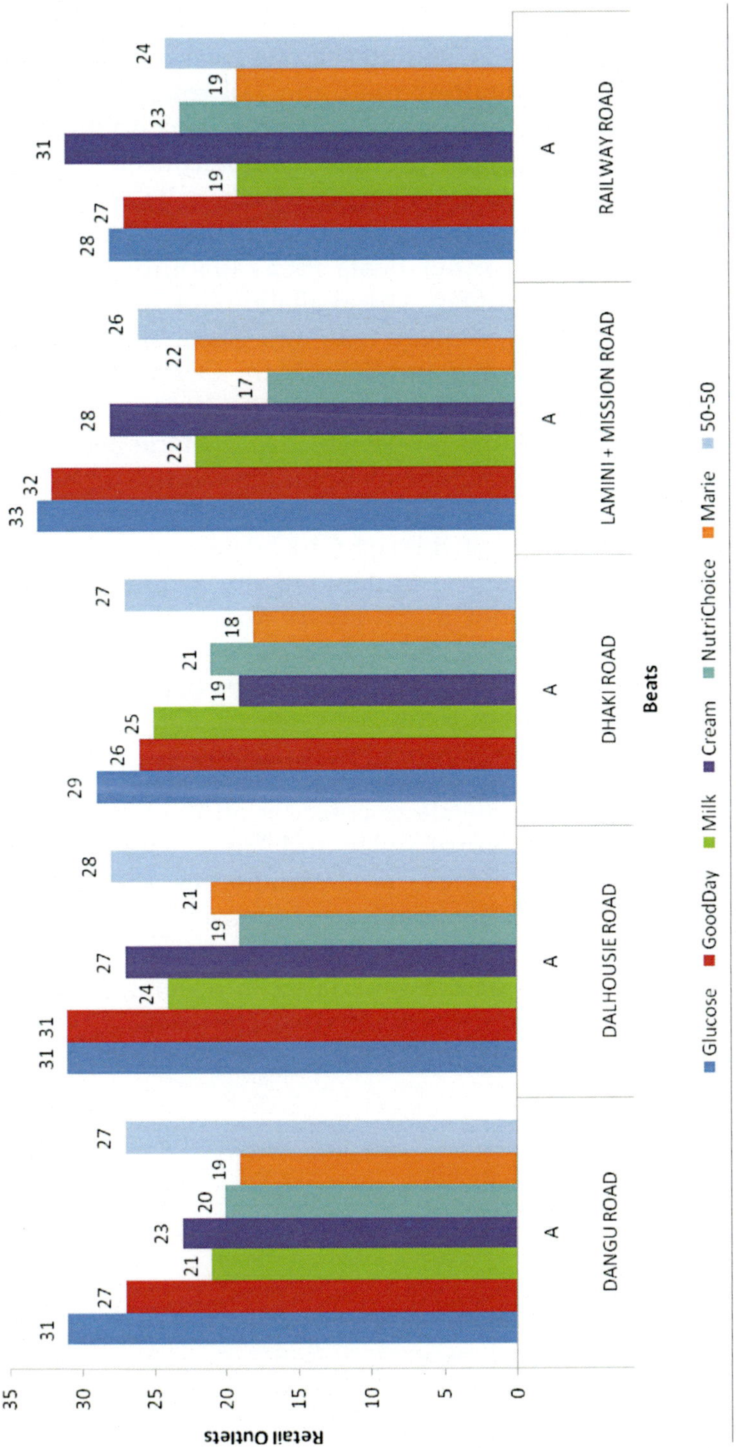

Figure 5. Availability of Britannia variants in Pathankot district.

From Figure 6, it has been analysed that also in district Gurdaspur, the variety of Glucose biscuits is leading the other variants in all the beats except the beats of Sangalpur road and Jail road. Glucose is followed by Cream biscuits in district Gurdaspur and holding the second position in terms of availability but the company needs to focus on it in the beats of College road and Babri Mani where its availability is less as compared to Glucose. In Sangalpur road beat GoodDay is leading other variants but in other beats its availability is less as compared to Glucose and GoodDay. The availability of 50-50 is good in the beats of Sangalpur road, College road and Geeta bhawan road but not so good in beats of Jail road and Babri Mani. Overall GoodDay and 50-50 biscuits are approximately equally available in district Gurdaspur. The variants of Milk, NutriChoice and Marie are found more unavailable than other variants of Britannia biscuits in all the beats of district Gurdaspur. So the company needs to focus mainly on these variants in all the beats of district Gurdaspur to make these variants of Britannia biscuits more available. Being a rural area, the customers of Gurdaspur are not much aware about these variants such as NutriChoice and Marie. The size of the biscuits and price according to size of biscuits also effects the availability of these variants which causes less demand by customers. So due to such reasons the unavailability of these variants is very high especially in rural areas.

Thus, at last the company needs to consider all the variants but needs to focus especially the NutriChoice and Marie variants in all beats of both districts as only these two variants are found more unavailable in both districts. So Britannia can increase the availability of these variants by giving more margins to retailers and also by introducing small size packs of these variants in order to increase their availability which in turn helps to increase the sales.

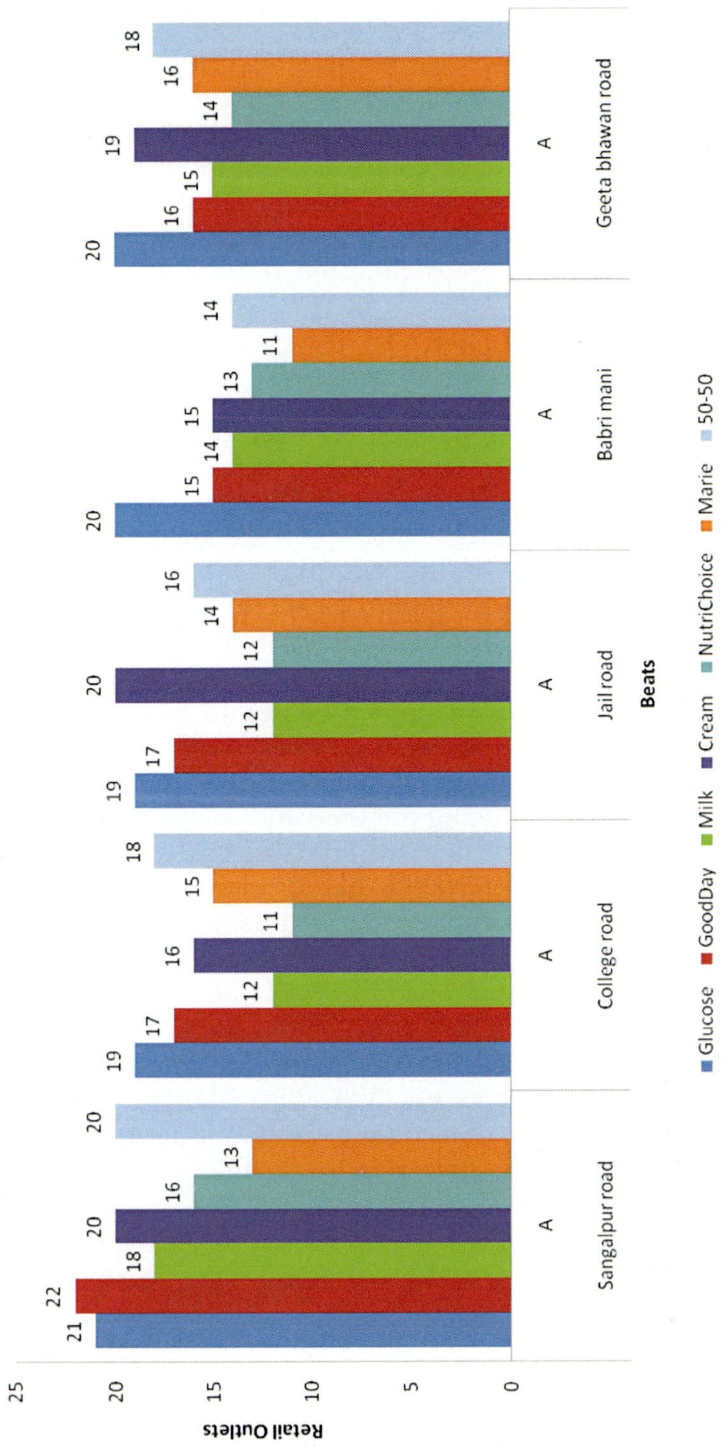

Figure 6. Availability of Britannia variants in Gurdaspur district

The KMO test provides the insight about the sample adequacy of the data for factor analysis. Whereas, the Barlett's test of sphericity reveals that whether there exist an correlation among the variables in the particular factor or not. Thus, after applying the factor analysis, the value of KMO from the results appears to be 0.518 which should be above 0.5 and also the value of Bartlett's test is statistically significant at 5% level of significance, so it can be assumed that the factor analysis test is suitable for the data/samples collected (Kaiser, H.F. (1963), Bartlett, M.S. (1950).

In the **Table 1**, all the variables having factor loading less than 0.45 are not shown, then the obtained results from the **Table 2**; shows that variables like Britannia is retaining the profitable customers only, Britannia is aware of the problems related to the retailers and Britannia is making the continuous or periodic visit to the retail outlets have highest loadings at first component hence they are grouped under the factor named focus on existing customers. Variables like Streamlining of business processes should be there, RTM should be launched and Britannia is taking much time in supplying the order placed by the retailer have highest loadings at second component so they are grouped under the factor named optimum utilization of resources. Variables like Britannia needs to reduce the operational cost, Automation of services is beneficial and Empowerment of distributor can be helpful in reducing the time and cost have highest loadings at third component so they are grouped under the factor called reduce operational cost. Variables like Britannia replaces the expiry products more quickly than their competitors, Britannia covers the breakage of products than other brands Sales distributor should be accessed with inventory data on daily basis and Britannia is providing incentives to those retailers who achieve their targets have highest loading at fourth component so they are grouped under the factor called optimize existing assets. Variables like Britannia is promoting its brands properly than others, Britannia is providing sound or better schemes and programs to the retailers, Britannia is properly implementing the Lakshya program and KAT program is beneficial for retailers have highest loading at fifth component hence they are grouped under the factor called maximize revenue opportunities.

Table 1. KMO and Bartlett's Test

Kaiser-Meyer-Olkin Measure of Sampling Adequacy.		0.518
Bartlett's Test of Sphericity	Approx. Chi-Square	591.356
	Df	190
	Sig.	.000

Note: tabular value of Chi-square at 5 per cent is 70.993

Table 2. Descriptive Statistics for Rotated Component Matrix.

Rotated Component Matrix

	Component				
	1	2	3	4	5
FEC1	**.620**				
FEC2	**-.593**				
FEC3	**.461**				
FEC4	-.372				
OUR1		**.607**			
OUR2		**.581**			
OUR3		**.475**			
OUR4		-.269			
ROC1			-.107		
ROC2			**-.636**		
ROC3			**.630**		
ROC4			**.549**		
OEA1				**.654**	
OEA2				**.650**	
OEA3				**.531**	
OEA4				**.511**	
MRO1					**.699**
MRO2					**.594**
MRO3					**.573**
MRO4					**.493**

Descriptive Statistics of the Variable

To carry out the analysis further, it becomes vital to calculate the descriptive statistics of all the variables calculated out after applying the Confirmatory factor analysis. The statistics help to determine the average response of the respondents against each factor. Consequently the measures of dispersion will provide an insight about the consistency and variation of the data towards the mean values.

The mean value of FOEC (2.936), OUR (3.339), ROC (3.728), OEA (3.555), MRO (3.320) and CRM (3.515) has been experienced from the **Table 3**. It has been observed that the measures of central tendency scores suggests that the average responses of the respondents appears to be on satisfied side of the 5 points likert scale.

In addition to this, the standard deviation, sample variance and coefficient of variation of the factor appears to be FOEC (0.654, 0.428, 22.282); OUR (0.487, 0.238, 14.5987); ROC (0.938, 0.879, 25.148); OEA (0.691, 0.478, 19.443); MRO (0.747, 0.559, 22.51); CRM (0.645, 0.416, 18.341) respectively. These statistics suggests that there is consistency in the nature of response by the respondents towards various factors.

Regression Analysis between CRM and its Strategies

Now we will apply regression analysis between customer relationship management and strategies of customer relationship management to know how much these strategies are affecting the customer relationship management.

$$CRM = \beta_0 + \beta_1 (FOEC) + \beta_2 (ROC) + \beta_3 (OUR) + \beta_4 (OEA) + \beta_5 (MRO)$$

Where

FOEC = (Britannia is retaining the profitable customers only, Britannia is aware of the problems related to the retailers, Britannia is making the continuous or periodic visit to the retail outlets).

ROC = (Streamlining of business processes should be there, RTM should be launched and Britannia is taking much time in supplying the order placed by the retailer).

OUR = (Britannia needs to reduce the operational cost, Automation of services is beneficial and Empowerment of distributor can be helpful in reducing the time and cost).

OEA = (Britannia replaces the expiry products more quickly than their competitors, Britannia covers the breakage of products than other brands Sales distributor should be accessed with inventory data on daily basis and Britannia is providing incentives to those retailers who achieve their targets).

MRO = (Britannia is promoting its brands properly than others, Britannia is providing sound or better schemes and programs to the retailers, Britannia is properly implementing the Lakshya program and KAT program is beneficial for retailers).

$\beta_{1,}$ β_2, β_3, β_4 and β_5 represent the regression coefficient of the corresponding factors.

A regression analysis is done to explain the variation in one variable (dependent variable), based on variation in one or more other variables (independent variables). In case there is one independent variable to explain the variation in one dependent variable, it is known as simple regression. If there are multiple independent variables to explain the variation in single dependent variables, it is known as a multiple regression model.

The value of correlation coefficient from **Table 4** appears to be 0.746 indicating that the independent variables show a high level of correlation with the dependent variable. The measure of strength of association in the regression analysis is given by the coefficient of determination denoted by R^2. This coefficient varies between 0 and 1 and represents the proportion of total variation in the dependent variable that is accounted for by the variation in the factors. From **Table 4**, the R^2 value is 0.557 which shows that 55.7 per cent of the variation in CRM can be explained by the five factors or independent variables in the given analysis.

To test the validity of the model, the study incorporated the ANOVA or F test as shown in **Table 5**. It can be depicted from the statistics that the corresponding F-value appears to be 41.52, which further appears to be statistically significant at 5 per cent level of significance. Thus the test statistics conclude the overall goodness of fit of the model.

Table 3. Descriptive statistics of the variable

	FOEC	*OUR*	*ROC*	*OEA*	*MRO*	*CRM*
Mean	2.936	3.339	3.728	3.555	3.320	3.515
Median	3.000	3.250	3.750	3.750	3.500	3.600
Mode	3.000	3.500	4.250	4.000	4.000	4.000
Standard Deviation	0.654	0.487	0.938	0.691	0.747	0.645
Sample Variance	0.428	0.238	0.879	0.478	0.559	0.416
Coefficient of Variation	22.282	14.5987	25.148	19.443	22.51	18.341

Table 4. Descriptive Statistics of Regression Analysis for CRM

Model Summary

Model	R	R Square	Adjusted R Square	Std. Error of the Estimate
1	0.746[a]	0.557	0.44	1.03321

Table 5. Analysis of Variance for Model Fitness

ANOVA

Model	Sum of Squares	Df	Mean Square	F	Sig.
1 Regression	1286.1	5	257.22	41.525	.000[a]
	Residual	1821.137	294	6.194	
	Total	3107.237	299		

Note: Tabular value at 5 per cent = 3.841From the Table 5.6, the equation can be written as:

CRM = -1.800 + 0.171(FOEC) + 0.940(OUR) + 0.087(ROC) + 0.226(OEA) + 0.308(MRO)

The regression model from the equation has been used to interpret the given result. From the equation it can be inferred that if focus on existing customer is increased by 1 unit, CRM is estimated to increase by 0.171 assuming all other variables to be constant. It is further been inferred that if optimum utilization of resources is increased by 1 unit, CRM is estimated to increase by 0.940 assuming all other variables constant. Coming to third factor if reduce operational cost is increased by 1 unit, it is estimated that there will be 0.087 increase in CRM assuming all other variables to be constant. Further if optimize existing assets is increased by 1 unit; CRM is estimated to increase by 0.226 assuming all other variables to be constant. At last it is inferred that if maximize revenue opportunities is increased by 1 unit, CRM is estimated to increase by 0.308 assuming all other variables to be constant.

We also have the ***t*-test** value for the significance of individual independent variables in the estimated analysis. From the **Table 6,** we can see that as there have been corresponding *t* values for the relevant coefficients of the estimated equation. These *t* values are used to validate the significance of the associated coefficients for the analysed results. It can be depicted from the table that the coefficient values of all the independent variables that are strategies of CRM appears to be statistically significant at 5 per cent level of significance as the estimated *t* values are greater than the tabular values, thus it can be concluded that all the coefficient values are statistically significant. Furthermore the p values from the table also confirm for the significance of estimated *t* values.

Table 6. Summary statistics of the Model

Coefficients

Model		Unstandardized Coefficients	Standardized Coefficients		T	Sig.
		B	Std. Error		Beta	
1	(Constant)	-1.800	1.477		-1.219	.224
	FOEC	.171	.064	.122	2.670	.008
	OUR	.940	.081	.526	11.552	.000
	ROC	.087	.042	.094	2.086	.038
	OEA	.226	.057	.194	4.001	.000
	MRO	.308	.053	.286	5.850	.000

Regression Analysis between Retailers Attitude and CRM

Now we will apply regression analysis between retailer's attitude and customer relationship management to know how much customer relationship management is affecting the retailer's attitude.

Retailers attitude = $\beta_0 + \beta_1$ (CRM)

Where

CRM = f (I have tried the brand and would use again but have not done so in the past 3 months, I have used the brand in the past 3 months but am not a regular user, I am a regular user but this is not my most often brand, I use this brand most often even though I do use other brands and I always use this brand as long as it is available).

β_1 represent the regression coefficient of the corresponding factor.

First of all the value of correlation from **Table 7** appears to be 0.737 indicating that the independent variables show a high level of correlation with the dependent variable. The measure of strength of association in the regression analysis is given by the coefficient of determination denoted by R^2. This coefficient varies between 0 and 1 and represents the proportion of total variation in the dependent variable that is accounted for by the variation in the factors. From Table 7, the R^2 value is 0.543 which shows that 54.3 per cent of the variation in RA can be explained by the CRM or independent variable in the given analysis.

To test the validity of the model, the study incorporated the ANOVA or F test as shown in Table 8. It can be depicted from the statistics that the corresponding F-value appears to be 105.95, which further appears to be statistically significant at 5 per cent level of significance. Thus the test statistics conclude the overall goodness of fit of the model.

RA= 4.703 + 0.310 (CRM)

The regression model from the equation has been used to interpret the given result. From the equation it can be inferred that if customer relationship management is increased by 1 unit, retailers' attitude is estimated to increase by 0.310 units.

Table 7. Descriptive Statistics of Regression Analysis for Retailers attitude

Regression Statistics	
Multiple R	0.737
R Square	0.543
Adjusted R Square	0.560

Table 8. Analysis of Variance for Model Fitness

ANOVA					
	Df	*SS*	*MS*	*F*	*Significance F*
Regression	1	298.748	298.748	105.959	0.000
Residual	298	840.199	2.819		
Total	299	1138.947			

Table 9. Summary statistics of the Model

	Coefficients	*Standard Error*	*t Stat*	*P-value*
Intercept	4.703	0.538	8.738	0.000
CRM	0.310	0.030	10.294	0.000

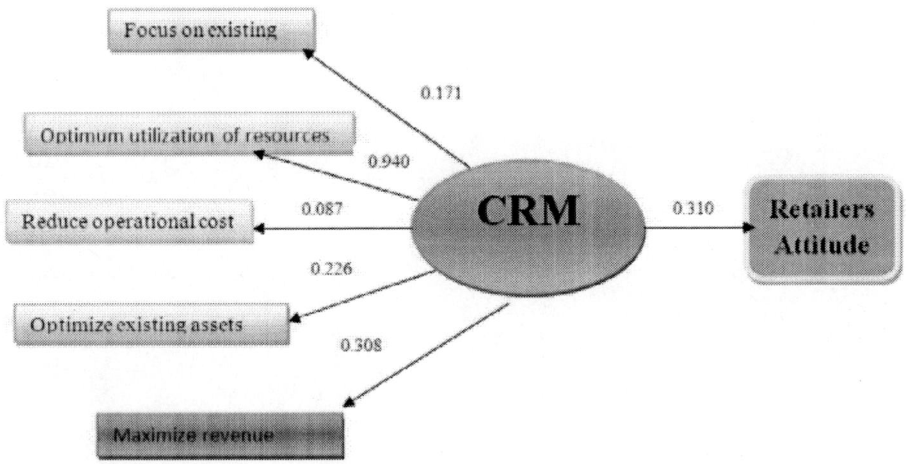

Figure 7. The following model is figured out from the analysis while applying the factor and regression analysis

While applying the regression analysis it has been identified that optimum utilization of resources strategy appears to be the vital variable for improving the customer relationship management at retail outlets in Pathankot and Gurdaspur districts of Punjab.

□□□

Chapter 3

Conclusion

It has been concluded from the study that Britannia is having more coverage in Pathankot district as compared to Gurdaspur district. However, the probable reason for the less coverage in Gurdaspur District is due to the presence of more rural population than in Pathankot. Moreover, the analysis conclude that there is a less number of retail outlets available in Gurdaspur District than in Pathankot leading towards decrease in market coverage and in particularly to the sales of Britannia biscuits.It has been analyzed that Britannia is giving tough competition to their competitors in all the beats except Dangu road where the position of Britannia to its major competitor Cremica appears to be on second position.While revealing the impact of various strategies on the customer relationship management it has been concluded that optimum utilization of resources appears to be the essential factor for CRM. Thus Britannia need to focus on Streamlining of business processes, launch of RTM and the actual products demanded by the retailer so as to focus on improving CRM. While analyzing the impact of CRM on retailer's attitude it has been revealed that the focus on improving the customer relationship management can improve the retailer's attitude towards the purchase of Britannia biscuits significantly.

Findings

It has been found that Britannia is having more coverage in Pathankot as compared to Gurdaspur because the distribution system of Britannia is not effective in Gurdaspur as compared to Pathankot as the large numbers of retailers are getting the product from local wholesaler due to high margins given on Britannia products along with the products of other local brands.

In district Pathankot, Britannia is having the highest coverage in Lamini and misson road beat, 33 out of 35 and lowest coverage in the beats of Dalhousie road and Railway road in which both beats are having 31 outlets covered out of 35. Whereas in district Gurdaspur, the highest coverage is in Sangalpur road beat with coverage of 22 retail outlets out of 25 and lowest coverage is in College road beat which is having the coverage of just 19 outlets out of 25.

In addition to this, it has been analysed from the statistics that Britannia is having more coverage in Pathankot district as compared to Gurdaspur district. However the probable reason for the less coverage in Gurdaspur district is due to the presence of more rural population then in Pathankot. Moreover the analysis conclude that there is a less number of retail outlets available in Gurdaspur district than in Pathankot leading towards decrease in market coverage and in particularly to the sales of Britannia biscuits. However, one of the most essential reasons for the less coverage and more non availability of Britannia biscuits in Gurdaspur district is due to the strong holding and presence of local brands especially Cremica. As these local brands were having less cost and more margin for the retailers in different beats of Gurdaspur district than in Pathankot. This has lead towards the presence of more local brands availability for biscuits than Britannia in Gurdaspur.

However, it has been found that one of the most essential reasons for the less coverage and more non availability of Britannia biscuits in Gurdaspur district is due to the strong holding and presence of local brands especially Cremica because of high margins given to retailers over the products of Britannia.

Marie and NutriChoice the strong variant in the fibre category of Britannia biscuits have shown low sale due to its high prices as compared to other variants of competing segments & found almost unavailable in all the beats of Pathankot & Gurdaspur.

Study also revealed the strong need for improving optimum utilization of resources strategy for improving the customer relationship management at retail outlets by conducting dealers meet and other related activities on regular intervals.

Positive retailer's attitude has been observed towards Britannia products. However due to lack of proper distribution and irregular visits to the retail outlets by the distributors, along with company representatives the retailers are forced to sell the products of Parle and ITC.

It is observed that most of the retailers of both the districts have demanded the automation of the services through RTM program by which the retailers will be able to access all the necessary information regarding the products, availability of stocks of particular variant of biscuits, margins on the products and schemes given by the company.

Policy Implications

It has been suggested that Britannia should focus on strengthening its existing distribution network in Gurdaspur as the local wholesaler is having the dominant grip over the market and it is attracting the existing retailers or prospects through lucrative offers given to them. The distributor of Britannia can also allow credit facilities to some extent in order to cover more market. Britannia needs to focus on the beats having less coverage in both districts. In Pathankot, the focus should be on the beats of Dalhousie road and Railway road. While in Gurdaspur, the beat that especially should be focused is College road beat. To retain its market leadership Britannia can increase their market share through sales promotion activities, intensive distribution and attractive schemes to retailers. Britannia should focus more on strengthening its supply chain. The size of the biscuit packs plays an important role in increasing the sales especially in rural areas. So Britannia should focus on the modification in the packaging of its biscuits mainly in the variants of Marie and NutriChoice. RTM should be launched in all the beats of district Pathankot and district Gurdaspur as soon as possible because it will help in reducing the communication gap between the company and the retailers and thereby improve the CRM.The delivery on specified date should be done by distributors when order is placed. A Grievance handling center should be instituted for handling retailers grievances related to Britannia products, complaints and suggestions by the retailers regarding the distributor's services if any. In addition to this, special programs can be organized to motivate the distributors and retailers to get awareness about the various schemes and also to push the sales. Britannia is found to be preferred by majority of the customers so point of purchases display through KAT program should be increased.

❑❑❑

Chapter 4

Bibliography

Aaker, D. A. and Keller, K. L., 1990. Consumer evaluations of brand extensions. *Journal of Marketing,* **54**: 27-41.

Ailawadi, K. L., Neslin, S. A., & Gedenk, K. 2001. Pursuing the value-conscious consumer: Store brands versus national brand promotions. *Journal of Marketing,* **65**: 71-89.

Anonymous. 2012. Biscuit Industry in India an Overview. *Federation of Biscuit Manufacturer's of India.* pp. 1-9.

Baker, W., Hutchinson, J. Wesley, Moore, D. and Nedungadi, P. 1986. "Brand familiarity and advertising: effects on the evoked set and brand preferences". In: advances in consumer research, 13. (Ed.) Lutz, R. J. (UT), *Association for Consumer Research,* pp. 146-147.

Barron, J. 2003. How strong brands get on intent and stay there. *Journal of Business Strategy,* **3**: 36-41.

Bello, Daniel and Gilliland. 1997. The effect of output controls, process controls, and flexibility on export market performance. *Journal of Marketing,* **61**: 22-38.

Best, R.J. 2005. Market-based management: strategies for growing customer value and profitability. 4th edition. Pearson: Prentice Hall.

Black, N. J., Lockett, A., Ennew C., Winklhofer H. and McKechnie S. 2002. Modelling consumer choice of distribution channels: An illustration from financial services. *The international Journal of Bank Marketing,* **20**: 161-173.

Chaudhuri, A. and M.B. Holbrook, 2001. "The chain of effects from brand trust and brand affect to brand performance: The role of brand loyalty," *Journal of Marketing,* **65**: 81-93.

Chen, Q. and H. Chen, 2003. Exploring the success factors of eCRM strategies in practice. *Database Marketing and Customer Strategy Management*, **11**: 333-343.

Coelho, F., and Easingwood, C. 2003. An exploratory study into the drivers of channel change. *European Journal of Marketing*, **42**: 1005-1022.

Corstjens, M. And Lal, R. 2000. "Building store and loyalty through private labels", *Journals of Marketing Research*, **37**: 281-291.

Corstjens, M. and P. Doyle 1979. "Channel optimization in complex marketing systems," *Management Science*, **25**: 1014-1025.

Cox, E. 2011. Retail analytics: the secret weapon. *International Journal of Auditing* **1**: 13-28.

Crosby, L.A. and J.R. Taylor, "Psychological commitment and its effects on post-decision evaluation".

Cunningham, I.C. M., Hardy, A.P., & Imperia, G. 1982. Generic brands versus national brands and store brands. *Journal of Advertising Research*, **22**: 25-32.

Czellar, S., 2003. Consumer attitude toward brand extensions: an integrative model and research propositions. *International Journal of Research in Marketing* **20**: 97-115.

Davis, S. 2002. Brand asset management: how businesses can profit from the power of brand. *Journal of Consumer Marketing*, **19**: 351-358.

De Chernatony, L. 1996. Integrated brand building using brand taxanomies. *Marketing Intelligence and Planning*, **14**: 40-45.

Dunner, David, & Narasimham, Chakravarthi. 1997. "The new appeal of private labels". *Harvard Business Review*, **77:** 45-65.

Elena Delgado *et al.* 2001. Brand trust in the context of consumer loyalty. *European Journal of Marketing*, **35**: 1238-1258.

Erdem, T., Zhao, Y., & Valenzuela, A. 2004. Performance of store brands: a cross-country analysis of consumer store-brand preferences, perceptions, and risk. JMR, *Journal of Marketing Research*, **41:** 86-100.

Farquhar, P. 1989. Managing brand equity, *Marketing Research*, **1**: 24-33.

Frazier, G. L. and Sheth, J., N, 1985. An attitude-behavior framework for distribution channel management. *Journal of Marketing*, **49**: 38-48.

Frazier, Gary 1985. "On the measurement of interfirm power in channels of distribution," *Journal of Marketing Research*, **20**: 254-62.

Frazier, Gary. 1983. On the measurement of inter firm: power in channels of distribution. *Journal of Marketing Research*, **20**: 158-166.

Grime, I., Diamantopoulos, A. and Smith, G., 2002. Consumer evaluations of extensions and their effects on the core brand. *European Journal of Marketing*, **36** : 1415-1438.

Hossein and Zakaria 2012. CRM benefits for customers. *International Journal of Engineering Research and Applications*. **2**: 1578-1586.

Injazz J. *et al*., 2003. Understanding customer relationship management: people, process and technology. *Business Process Management Journal*. **19**: 672-688.

John, Pracejus, G. Douglas Olsen. 2004. The role of brand fit in the effectiveness of cause-related marketing campaigns. *Journal of Business Research*, **57**: 635-640.

Keller, K.L. 1993. "Conceptualizing, measuring, and managing customer-based brand equity", *The Journal of Marketing*, **57**: pp.1.

Keller, Kevin Lane 2000. "The brand report card". *Harvard Business Review*, **78**: 147-157.

Keller, Kevin Lane and Aaker, David A. 1997. "Managing the corporate brand: the effect of corporate marketing activity on consumer evaluations of brand extensions", Working Paper Report No. 97-106, Cambridge, MA: Marketing Science Institute.

Kotler, P. 2003. Marketing management: The effects of perceived interdependence on dealer attitudes. *Journal of Marketing Research*, **32**: 348-356.

Kotler. P., and Keller, K. 2009. Marketing management, 13th edition: Prentice Hall New Jersey.

Kumar, V. and Reinartz, W. 2012. Impact of CRM on marketing channels, CRM, springer text in business and economics, pp. 235-260.

Lassar, W., B. Mittal, and A. Sharma, 1995. "Measuring Customer-Based Brand Equity," *Journal of Consumer Marketing*, **12** :11-19.

Merrie Brucks, *et al.* 2000. Price and brand name as indicators of quality dimensions for consumer durables. *Journal of the Academy of Marketing Science*. **28**: 359-374.

Morgan, R.M. and S.D. Hunt, 1994. "The commitment-trust theory of relationship marketing." *Journal of Marketing*, **58:** 20-38.

Morgan, Robert M., and Shelby D. Hunt 1994. The commitment-trust theory of relationship marketing. *Journal of Marketing*, **58**: 20–38.

Mueller, B. 2010. Dynamics of international advertising: theoretical and practical perspectives. *Journal of Consumer Research*, **29**: 77-90.

Nandan, S. 2005. An exploration of the brand identity-brand image linkage: A communications perspective, *Journal of Brand Management*, **12** : 264-278.

Neuwirth B. 2012. Marketing channel strategies in rural emerging markets. *Unlocking Business Potential* **7**: 01-40.

Obaji, R.N. 2011. The effects of channels of distribution on Nigerian product sales. *The International Business and Economics Research Journal*, **8**: 85-91.

Payne, A. and Frow, P. 2005. A Strategic Framework for Customer Relationship Management, *Journal of Marketing*, **69**: 167-191.

Payne, A., Christopher, M., Clark, M., Peck, H. 1999. ''Relationship marketing for competitive advantage'', Butterworth Heinemann, Oxford.

Pelsmacker D, Geuens, M. and Van den Bergh, J. 2004. Marketing communications: *A European perspective*. FT Prentice Hall, Harlow.

Peppers, D. and Rogers, M. 2011. Managing customer relationships: a strategic framework. *Indian Journal of Research*, **3**: 30-34.

Ramanathan. 2007. Retailing channel enhancement strategies adopted by FMCG companies in south Indian rural markets. *The Icfaian Journal of Management Research*, **6**: 66-70.

Reinartz, W., Krafft, M., and Hoyer, W.D. 2004. The customer relationship management process: its measurement and impact on performance, *Journal of Marketing Research* **41**: 293- 313.

Rosenbloom, B. 2004. Marketing channels: *A Management View*, 7th edition. South-Western of Thomson.

Rust, R.T., Ambler, T., Carpenter, G. S., Kumar V. & Srivastava, R. K. 2004. Measuring marketing productivity: current knowledge and future directions, *Journal of Marketing*, **68**: 76-89.

Sayman, S., & Raju, J. S. 2004. Investigating the cross-category effects of store brands. *Review of Industrial Organization*, **24**: 129-141.

Speed, R., 1998. Choosing between line extensions and second brands: the case of the Australian and New Zealand wine industries. *Journal of Product and Brand Management* **7**: 519-536.

Srivastava, R.K. and Shocker, A.D., 1991. Brand equity: a perspective on its meaning and measurement. Working Paper 91-124, *Marketing Science Institute,* Cambridge, MA.

Stiener, Robert L. 1993. "The inverse associating between the margins of manufacturers and retailers" *Review of Industrial Organization,* 717-740.

Webster, Frederick E., Jr. 1976, "The role of the industrial distributor in marketing strategy," *Journal of Marketing*, **40**: 10-16.

Wileman, A., Jary .M. 1997. "Retail power plays: from trading to brand leadership", *Macmillan Business, Great Britian,* 45-65.

Wood, L. 2000, Brands and brand equity: definition and management. *Management Decision*, **38**: 662-669.

Yao, H.I, and Khong, K.W. 2011. Effectiveness of customer relationship management on customer satisfaction in the commercial banks of Taiwan. *Cotemporary Management Research*, **7**: 105-106.

Yoo, B., Donthu, N. & Lee, S. 2000. An examination of selected marketing mix elements and brand equity. *Academy of Marketing Science Journal*, **28**: 195-211.

About the Authors

Mr. Sajad Khurshid Khan possesses a meritorious academic record and awarded by university merit certificate in recognition of his having achieved the prescribed scholastic standards of high order OGPA in Masters (Agri-Business Management) programme. He has obtained his degree in Agricultural Business Management from the Sher E Kashmir University of Agricultural Sciences and Technology of Jammu. He has 1 research publication to his credit.

Dr. Jyoti Kachroo has remained gold medalist (Two) and meritorious throughout her academic career, graduated and post graduate (Economics) from University of Jammu, Jammu and Kashmir. She was initially appointed as lecturer (Economics) in Govt. Degree Collage Jammu and later joined the division of agricultural Economics and Statistics, SKUAST Jammu as Associate Professor. Presently she is professor and Head, Division of Agricultural Economics & Agri-Business Management. Her field of expertise is Women empowerment, agricultural production economics, marketing and econometric model based research. She has published about 85 articles which include chapters, popular articles, review papers and about 40 research publications in referred journals. She has been honored with D.T.Doshi Foundation award for best presentation. She has 24 years experience of teaching and research and she has guided 16 research scholars.

Mr. Naveed Hamid possesses a meritorious academic record and awarded by university merit certificate in recognition of his having achieved the prescribed scholastic standards of high order OGPA in Masters (Agri-Business Management) programme. He has obtained his degree in Agricultural Business Management from the Sher e Kashmir University of Agricultural Sciences and Technology of Jammu. He has 26 research publications in referred and UGC journals, 10 abstracts, 02 books and 4 international conferences to his credit. He is member and reviewer of various agri-business, economics and management journals. He is awardee of a prestigious project by NABARD in 2014 and three award from repute organization for his contribution to the multidisciplinary field. His field of expertise is Agricultural Business Management, Supply chain/value chain and marketing. Mr. Naveed Hamid is presently working as Assistant Professor ©, Agri-Business Management in Division of Agricultural Economics & ABM, SKUAST Jammu. In his two and a half year experience in teaching and research, he has guided 5 research PG Scholars.